Lecture Notes in Mathematics

Edited by A. Dold and B. Eckmann

893

Geometries and Groups

Proceedings of a Colloquium Held at the
Freie Universität Berlin, May 1981

Edited by M. Aigner and D. Jungnickel

Springer-Verlag
Berlin Heidelberg New York 1981

Editors

Martin Aigner
2. Mathematisches Institut der Freien Universität Berlin
Königin-Luise-Str. 24–26, 1000 Berlin 33,
Federal Republic of Germany

Dieter Jungnickel
Mathematisches Institut der Universität Gießen
Arndtstr. 2, 6300 Gießen, Federal Republic of Germany

AMS Subject Classifications (1980): 05-06, 20-06, 51-06

ISBN 3-540-11166-2 Springer-Verlag Berlin Heidelberg New York
ISBN 0-387-11166-2 Springer-Verlag New York Heidelberg Berlin

Printed in Germany

Printing and binding: Beltz Offsetdruck, Hemsbach/Bergstr.
2141/3140-543210

Dedicated to

Professor Dr. Hanfried Lenz

on the occasion of his 65th birthday

This volume contains the proceedings of a colloquium in honour of Prof. Dr. Hanfried Lenz, organized by the Freie Universität Berlin in May 1981 to mark both Prof. Lenz's 65th birthday and his retirement (from formal duties, not from mathematics !). Though Prof. Lenz worked in many areas of mathematics, his main interest has been in geometry and - for about the last 10 years - in particular in finite geometries and designs (for more details on his work see the following "Geleitwort"). It was decided to focus attention on the combinatorial and group theoretic aspects of geometry. Five survey lectures (of two hours each) were invited (given by F. Buekenhout, J. Doyen, D.R. Hughes, U. Ott and K. Strambach); the corresponding papers constitute the first part of this volume. There were also about 30 contributed talks (which were not restricted to the area mentioned above); 11 contributed papers fitting into this area form the second part of this volume which will hopefully be of interest to anyone working in geometries and groups.

We finally have one more remark: it was a great pleasure to be able to present Prof. Lenz with an ingenious machine (designed and built by Th. Beth and W. Fumy of the Universität Erlangen) which visibly displays some of the applications of combinatorial theory. The article by Fumy deals with the mathematical background and the powers of this delightful device.

Berlin and Gießen, July 1981

M. Aigner

D. Jungnickel

TABLE OF CONTENTS

GELEITWORT

Dieser Band ist Hanfried Lenz zu seinem 65. Geburtstag gewidmet. Daher wollen wir uns hier sein bisheriges mathematisches Werk vor Augen führen, was bei der großen Zahl seiner Veröffentlichungen allerdings nur in Auswahl geschehen kann. Ein solches Auswählen ist schwer, da Hanfried Lenz auf vielen verschiedenen Gebieten der Mathematik gearbeitet hat und ein Abwägen der Bedeutung immer subjektiv sein wird.

Am Anfang seines mathematischen Schaffens stehen einige kleinere Arbeiten zur Analysis und Funktionentheorie (1951/2), und auf diesen Problemkreis kommt er später (1956/7) noch einmal zurück. Aber schon 1952 wendet sich sein Interesse der Geometrie zu und zwar zuerst der Theorie der projektiven Ebenen, die eigentlich erst seit den vierziger Jahren ein Eigenleben gewann. 1953 entdeckt er das erste Beispiel einer endlichen projektiven Ebene , in der einige, aber nicht alle vollständige Vierecke kollineare Diagonalpunkte haben[1]. 1954 erscheint seine Arbeit "Kleiner Desarguesscher Satz und Dualität in projektiven Ebenen"[2], in der er die projektiven Ebenen nach der Menge der Paare (P,g) (P Punkt auf Gerade g) klassifiziert, für welche die Gruppe der (P,g)-Kollineationen (alle Geraden durch P und alle Punkte auf g bleiben fest) transitiv auf den Punkten $\neq$P einer Geraden $\neq$g durch P wirkt. Diese Klasseneinteilung wurde 1957 durch Barlotti verfeinert, indem er auch die nichtinzidenten Paare (P,g) einbezog. Als Lenz-Barlotti-Klassifikation bezeichnet gehört diese Einteilung seitdem zum Standardwerkzeug in der Theorie der projektiven Ebenen. Ebenfalls 1954 erscheint eine umfangreiche Arbeit "Zur Begründung der analytischen Geometrie"; hier (und in einer im gleichen Band der Sitzungsbericht der Bayerischen Akademie der Wissenschaften erschienen kleineren Note) wird die projektive Geometrie beliebiger (auch unendlicher) Dimension ≥ 3 aus Verknüpfungsaxiomen entwickelt und ebenso die affine Geometrie unter Benutzung eines Parallelismus in der Geradenmenge. Ebenso geht Hanfried Lenz 1958 in seinem mathematikdidaktisch orientierten Beitrag "Ein kurzer Weg zur analytischen Geometrie"[2] vor. Hier weist er auf die von Artin, Baer, Dieudonné verwendete koordinatenfreie Schreibweise hin ("an Einfachheit und Allgemeinheit der älteren formalen Behandlung der linearen Algebra überlegen") und bemerkt in einer

---

1) Arch.Math. 4, 327-330
2) Jahresber. DMV 57, 20-31
3) Math.Phys.Semesterber. 6, 57-67

Fußnote (hier als in ihrer Formulierung für den Verfasser typisch zitiert): "Die unter Fußnote 6 zitierten Arbeiten Lenz 1953/4 geben in dieser Hinsicht Beispiele, wie man es nicht machen soll". Auch seine 1961 bzw. 1965 erschienenen Bücher "Grundlagen der Elementarmathematik" (3.Aufl. 1975) und "Vorlesungen über projektive Geometrie" gehen teilweise auf diese Beschäftigung mit den Grundlagen der projektiven und affinen Geometrie zurück. Beide Werke enthalten viel bemerkenswert Neues, u.a. über quadratische Formen. Im zweiten dieser Bücher heißt es über die Bedeutung der projektiven Geometrie: "Wenn man von der z.Zt. blühenden Erforschung der projektiven Ebenen absieht, bietet die projektive Geometrie heute nicht noch viele offene Fragen: Sie ist aber unentbehrlich 1. zur Zusammenfassung vieler klassischer geometrischer Theorien im Sinne des Erlanger Programms von Felix Klein, 2. für die Begründung der nichteuklidischen Geometrie und 3. als Vorstufe für das Studium des sehr ausgedehnten und schwierigen Fachgebietes der algebraischen Geometrie. Für diese Zwecke erkennt auch Dieudonné die Existenzberechtigung der projektiven Geometrie an, die er im übrigen veraltet nennt. Da er selbst schöne Beiträge zur projektiven Geometrie geliefert hat, sollte man m.E. diese Kritik nicht tragisch nehmen, sondern vielmehr die vielen bewährten Schlußweisen und Ergebnisse der projektiven Geometrie in den modernen Aufbau der Mathematik einbauen."

Aus den Jahren 58, 59 sind drei Arbeiten[4)] hervorzuheben, die Beiträge zum Helmholtzschen Raumproblem in endlichdimensionalen reellen bzw. komplexen Vektorräumen liefern. Dem Bemühen, den mathematischen Hintergrund des im Mathematikunterricht der Schule Gelehrten aufzuklären, das Hanfried Lenz schon zum Schreiben seiner "Grundlagen der Elementarmathematik" veranlaßt hatte, entspringt 1967 eine die ordnungstheoretischen Grundlagen aufklärende Begründung der Winkelmessung[5)]. Seine Stellung zu Fragen des Mathematik- und insbesondere des Geometrieunterrichts findet man eindringlich beschrieben in dem Beitrag "A bas Euclide - vive Bourbaki?" aus dem Jahr 1963:[6)] Man sollte "Euklid" durch "Bourbaki" erneuern und ergänzen, aber nicht ersetzen (leider wurde diese Mahnung in den folgenden Jahren zu wenig berücksichtigt); eine besonders bemerkenswerte von drei Thesen lautet: "Psychologie und Didaktik haben den Vorrang vor logischer Abstraktion und Axiomatik, sind aber damit verträglich."

---

4) Arch.Math. 8,477-480, Math.Ann.135,244-250, u. 137,150-166.
5) Math. Nachr. 33, 363-375
6) Praxis d.Math. 5, Heft 4, 85-87

1975/6 wendet sich Hanfried Lenz einem neuen Forschungsgebiet zu, der endlichen Geometrie: Zwei gemeinsam mit D. A. Drake geschriebene Arbeiten[7] bringen einen Durchbruch in der Existenzfrage für endliche Hjelmslev-Ebenen; 1977 beschreibt er[8] ein Konstruktionsverfahren für spezielle endliche Inzidenzstrukturen (mittels Einsetzen von Inzidenzmatrizen in Inzidenzmatrizen), das sich in der Folge in Arbeiten von Lenz und anderen Verfassern als sehr fruchtbar in der Theorie der Blockpläne, Steiner-Systeme und anderer Design-Klassen erweist. Gemeinsam mit Th. Beth und D. Jungnickel schreibt Hanfried Lenz an einem umfangreichen Buch "Design Theory" , das viele neue Ergebnisse von ihm und seinen Mitarbeitern enthalten wird. Für dieses Buch und für seine weiteren Forschungen auf dem Gebiet der "Designs" wünschen wir Hanfried Lenz auch in unserem Interesse viel Erfolg.

Im Namen aller Freunde, Mitarbeiter und Kollegen von Hanfried Lenz

Günter Pickert

---

7) Abh.Math.Sem. Hamburg <u>44</u>, 70-83; Bull.Amer.Math.Soc. <u>82</u>, 265-267

8) In: Beiträge zur Geometrischen Algebra; 225-235

THE BASIC DIAGRAM OF A GEOMETRY

Francis Buekenhout
Department of Mathematics CP.216
Université Libre de Bruxelles
1050 Bruxelles-Belgium

1. Introduction.

The language of diagrams for geometries has been devised by Tits in the 1950's and it has been developed by him toward a complete combinatorial-geometric theory of the simple groups of Lie-Chevalley type [27]. Recently the author observed that the basic ideas of the diagram language could be applied to geometries arising from some of the sporadic groups [1]. In this context inspiration came primarily from the deep investigation of groups with a class of 3-transpositions due to B.Fischer [8]. This discovery of geometries for the sporadic groups gave a new impulse to some of the earliest ideas of Tits. For an excellent historical introduction to these ideas we refer to Tits [29].

During the last few years, an important activity on diagram geometries has been growing and it seems likely that this trend is going to increase in the near future.
There are several reasons for this :
a) diagram geometry provides a unified setting for many geometric theories in a field where this is badly needed;
b) as a consequence, known theories and known methods are better understood;
c) as another consequence, an enormous potential of new problems and of variations on known problems or known results is easily made available;
d) relationships with other fields are made easier.
The last statement is particularly well illustrated by the following examples :
- geometric interpretation of the finite simple groups and possible influence on the "revision" of the classification theory for these groups (in the sense of Gorenstein)
- representation theory in the spirit of the Feit-Higman theorem which is the subject of Ott's report at this conference (see also Ott [16]).
- the obvious connection with combinatorial topology (see Ronan [20], [21], Tits [30]).

In view of these characteristics it is desirable as always to work on firm foundations. This is the purpose of our paper. In [1],[3] basic concepts and general theorems are exposed. Various persons have pointed out shortcommings of different kinds in these papers and a series of suggestions and improvements have been made. Altogether the evolution of foundations with respect to [1] is sufficient to justify a completely new synthesis.

The author would like to acknowledge the help of G.Glauberman and M.Perkel for their most careful reading of [1] and their numerous suggestions **and A. Brouwer who made several corrections in this paper.**

2. Geometries over a set.

Let Δ be a set which can be seen as a set of indexes, a set of names such as point, line, plane,...,a set of colors, etc. Here elements and subsets of Δ will be called types.

A geometry Γ over Δ is a triple $\Gamma=(S,I,t)$ where S is a set (the elements of Γ), I is a symmetric and reflexive relation defined on S (the incidence relation of Γ) and t is a mapping of S onto Δ (the type function of Γ) such that (TP) the restriction of t to every maximal set of pairwise incident elements is a bijection onto Δ (transversality property).

Comments. 1) We shall also use the notation $\Gamma_i=t^{-1}(i)$ for each $i \in \Delta$. In other words, Γ_i is the set of all elements of Γ of type i.

2) In [1] we would rather speak of incidence structures and varieties instead of geometries and elements. The present vocabulary has been adopted in most recent talks on the subjet, in particular [29].

3) Notice that a geometry can be seen as a multipartite graph with distinct colors on the distinct components.

4) One could equivalently see a geometry over Δ as a set endowed with a partition indexed by Δ or see it as a quadruple (S,I,Δ,t) etc.

5) The axioms numbered (1), (2), (3) in [1] and required there from an incidence structure will not be required from a general geometry with the exception of (TP). There is indeed evidence that this more general viewpoint may be technically useful as is pointed out by Percsy [18], [19].

6) Strangely enough the concept of geometry as presented here appears very clearly in a paper of E.H.Moore as early as 1896 [14].

We shall need some more definitions. Let $\Gamma=(S,I,t)$ be a geometry over Δ.

A flag F of Γ is a (possibly empty) set of pairwise incident elements of Γ. Two flags F_1,F_2 of Γ are called <u>incident</u> and we write $F_1 \; I \; F_2$ if $F_1 \cup F_2$ is still a flag. Maximal flags are called <u>chambers</u>. The set of all chambers of Γ is denoted by Cham Γ.

The <u>type</u> of a flag F is the set t(F). If A is a subset of Δ then a flag of type A will also be called an <u>A-flag</u>. If A is reduced to a single element of Δ then an A-flag is also called an <u>A-element</u>. Hence if $i \in \Delta$, then Γ_i is the set of all i-elements of Γ or the set of all elements of type i. The <u>cotype</u> $t^*(F)$ of a flag F is the set Δ-t(F).

The <u>residue</u> of a flag F in Γ is the geometry $\Gamma_F=(S_F,I_F,t_F)$ over $t(S_F)=\Delta-t(F)$ defined by : S_F is the set of all elements of Γ not in F, incident with all elements in F, I_F (resp.t_F) is the restriction of I (resp.t) to S_F. Clearly axiom (TP) holds.

2.1.<u>Proposition</u>. If $\Gamma=(S,I,t)$ is a geometry over Δ and F is a flag of Γ then the residue Γ_F is a geometry over Δ-t(F).

The <u>rank</u> $r(\Gamma)$ of a geometry Γ over Δ is the cardinality of Δ i.e. the number of distinct types in which the elements of Γ fall. If F is a flag then the <u>rank</u> r(F) of F is the cardinality of F and the <u>corank</u> of F is the rank of Γ_F.

Two chambers are called <u>adjacent</u> if their intersection is a flag of corank 1. This provides clearly a graph structure on Cham Γ.
Let $\Gamma=(S,I,t)$ be a geometry over Δ and $\Gamma'=(S',I',t')$ be a geometry over Δ' such that $S' \subseteq S$, $I' \subseteq I$, $\Delta' \subseteq \Delta$ and t' is the restriction of t to S'. Then Γ' is a <u>subgeometry</u> of Γ. If in addition I' is the restriction of I to S' then Γ' is an <u>induced</u> <u>subgeometry</u> of Γ.

If Δ is a set then the <u>unit geometry</u> 1_Δ over Δ is the geometry $\Gamma=(\Delta,I,t)$ where I consists of all pairs $(x,y) \in \Delta^2$ and t is the identity map.

<u>Comments</u> 7) A geometry Γ determines a flag complex $F(\Gamma)$ in the sense of Tits [27] whose elements are the flags of Γ and in which the order relation is inclusion. The notion of flag complex is however more general than that of geometry since any graph and not just a multipartite graph determines a flag complex.

8) In the context of flag complexes the residue of a flag F as defined here, corresponds to the <u>star</u> of A or St A [27] which is also called the <u>link</u> of A in topology.

9) In [30], Tits introduces another approach of geometries based on chambers which was first suggested by L.Puig. A <u>chamber system</u> over the set Δ consists of a set C whose elements are called <u>chambers</u> together with a system of partitions of C indexed by Δ.
Clearly every geometry Γ determines a chamber system which we shall

denote abusively by Cham Γ: for every $i \in \Delta$, two chambers are called <u>i-equivalent</u> if their intersection is of cotype i; the equivalence classes of this relation determine a partition of Cham Γ corresponding to i.

Conversely every chamber system C determines a geometry $\Gamma(C)$ (see [30]). In general Γ does not coincide with $\Gamma(\text{Cham } \Gamma)$; however this holds under fairly mild conditions [30] namely that Δ is finite and that Γ is strongly connected (see section 5).

10) A geometry of rank 0 consists of the empty set and the empty incidence relation.

Geometries of rank 1 consist trivially of any set and a uniquely determined incidence relation.

3. Morphisms.

Let $\Gamma=(S,I,t)$ and $\Gamma'=(S',I',t')$ be geometries over sets Δ and Δ' respectively.

A <u>morphism from Γ to Γ'</u> is a pair (α,β) where $\alpha:S \to S'$, $\beta:\Delta \to \Delta'$ are mappings such that x I y implies $\alpha(x)$ I' $\alpha(y)$ and $t' \circ \alpha = \beta \circ t$.

Equivalently a morphism is a mapping $\alpha:S \to S'$ such that α preserves incidence and type equality i.e. $t(x)=t(y)$ implies $t'(\alpha(x))=t'(\alpha(y))$.

A <u>Δ-morphism</u> is a morphism (α,β) such that β is the identity i.e. a morphism preserving types.

Isomorphisms, automorphisms, Δ-isomorphisms and Δ-automorphisms are defined in the standard usual way.

<u>Comments</u> 1) Important automorphisms of geometries which are not Δ-automorphisms are provided by dualities, polarities and trialities (see for instance [26]). Morphisms which are not isomorphisms tend to play an increasing role in geometry. This is the case of <u>foldings</u> in the theory of buildings [27] (see also [9], [31]).

Embedding a geometry into another geometry involves a morphism of course (see Percsy [18]).

The local theory of buildings developed by Tits [30] and the theory of universal covers of chamber systems as treated by Ronan [20] rely on a special kind of morphism or local isomorphism.

2) For every geometry Γ over Δ there is a canonical Δ-morphism on the unit geometry 1_Δ over Δ namely the type function t.

3) Under a morphism α the image of a flag F is a flag $\alpha(F)$ and the image $\alpha(\Gamma_F)$ is contained into $\alpha(\Gamma)_{\alpha(F)}$.

4) Here is a fundamental procedure leading to interesting subgeometries in various contexts. Let α be a morphism of Γ into Γ (endomorphism). Then the induced subgeometry Abs α(absolute of α) consists of all elements X of Γ such that $\alpha(X)$ I X. If α is a Δ-morphism then Abs α consists of all elements fixed by α. If α is a polarity of a projective space then Abs α is a polar space or better Abs α determines a covering of a polar space (see below). If α is a triality of a polar space of type D_4 then Abs α determines a covering of a generalized hexagon.

We shall have some use of the following definition given by Tits in [30]. A Δ-morphism α is a covering if α is surjective and if for every $x \in S$ the restriction of α to the residue Γ_x is an isomorphism onto $\alpha(\Gamma)_{\alpha(x)}$. If there is such a covering then we shall also say that Γ is a covering of Γ'.

4.Thickness.

A geometry Γ is firm (resp.thick) if every non-maximal flag F of Γ is contained in at least two (resp.three) chambers of Γ.
A geometry Γ is thin if every flag of corank 1 is contained in exactly two chambers of Γ.

4.1.Proposition. A geometry Γ is firm (resp.thick, thin) if and only if every rank 2 residue of (a flag in) Γ is firm (resp.thick, thin).
Proof. Straightforward.

Comments 1) Tilings, polyhedra and polytopes provide examples of thin geometries.
2) A geometry Γ over Δ is a covering of the unit geometry 1_Δ if and only if every non-empty flag of Γ is contained in exactly one chamber.

Let $\Gamma=(S,I,t)$ be a geometry over Δ and let $i \in \Delta$. Then the i-order of Γ is the set of all numbers n-1 where n is the number of chambers containing some flag of cotype i. If Γ_F is some residue of Γ with $i \in t(\Gamma_F)$ then the i-order of Γ_F is contained in the i-order of Γ.

A thin geometry is characterized by the property that all of its i-orders are equal to 1. A covering of the unit geometry is characterized by the property that all of its i-orders are equal to 0.

If the i-order of Γ is reduced to a single number for each i then Γ is called order regular. In that case every residue Γ_F is also order regular.

5. Connectedness.

We shall say that a geometry $\Gamma=(S,I,t)$ over Δ is strongly connected if for every distinct i,j in Δ, $t^{-1}(i) \cup t^{-1}(j)$ is a connected graph for the incidence relation and if the same property holds in every residue Γ_F where F is a flag of Γ.

Comments 1) This is axiom (2) in [1]. The terminology is that of Tits [29]. In [30], Tits uses a weaker notion of connectedness and a notion of residual connectedness which is equivalent to strong connectedness.
2) Notice that the empty geometry and any rank 1 geometry are strongly connected. From the point of view of graph theory or topology this is a good reason to distinguish strong connectedness and connectedness.
3) In [30] a geometry $\Gamma=(S,I,t)$ is called connected if the graph (S,I) is connected (which implies S non empty).

Every strongly connected geometry of rank $\geqslant 2$ is connected. Γ is called simply connected (resp. strongly simply connected) if it is connected (resp. strongly connected) and if every covering by a connected geometry is an isomorphism (resp. if all residues of flags of corank $\geqslant 3$ in Γ are simply connected).
4) Observe that strong connectedness does not imply firmness : the unit geometry over Δ provides a counter-example.
5) If Γ is strongly connected then every residue Γ_F is strongly connected.
6) Assume Γ is a firm geometry of finite rank. Then Γ is strongly connected if and only if the set Cham Γ provided with the adjacency relation is a connected graph and the same property holds for Cham Γ_F where F is any flag of Γ.
The proof of this property is fairly straightforward. It has not appeared fully in print. For a partial proof see A.Valette [31].

6. Direct sums of geometries.

The concept of direct sum is fully recognized and used in Tits [25]. For a more explicit and detailed study we refer to A.Valette [31] whose work will be closely followed here.

Let J be a set of indices and let $(\Delta_j)_{j \in J}$ be a family of sets.

For each $j \in J$ let $\Gamma_j=(S_j,I_j,t_j)$ be a geometry over Δ_j. We assume that the Δ_j's are pairwise disjoint as well as the S_j's.

6.1. <u>The direct sum of the geometries</u> Γ_j is the geometry $\Gamma = \bigoplus_{j\in J} \Gamma_j = (S,I,t)$ defined as follows:

1) Δ is the union of the Δ_j's ;
2) S is the union of the S_j's ;
3) $I|S_j=I_j$ and $x\ I\ y$ whenever x and y belong to dinstinct components Γ_j,Γ_k respectively ;
4) $t|S_j=t_j$.

<u>Examples</u> 1) A direct sum of rank 1 geometries is a complete multipartite graph and conversely. A direct sum of two rank 1 geometries is called a <u>generalized digon</u>. These rank 2 geometries play a fundamental role in the theory.
2) If Γ is a geometry over Δ and 0 denotes the empty geometry over the empty set then Γ is obviously isomorphic to $\Gamma \oplus 0$.
3) A unit geometry of rank n is the direct sum of n unit geometries of rank 1.

<u>Properties</u> (see A.Valette [31]).
6.1. F is a flag (resp.chamber) of $\bigoplus_{j\in J}\Gamma_j$ if and only if $F \cap S_j$ is a flag (resp.chamber) of Γ_j for every $j \in J$.
6.2. F is a flag of corank one of $\bigoplus_{j\in J}\Gamma_j$ if and only if there is a unique $j \in J$ such that $F \cap S_j$ is a flag of corank one of Γ_j and for $k\neq j$, $F \cap S_k$ is a chamber of Γ_k.
6.3. There is a canonical bijection from Cham $\bigoplus_{j\in J}\Gamma_j$ onto the cartesian product $\prod_{j\in J}$ Cham Γ_j which completely describes the adjacency relation on Cham $\bigoplus_{j\in J}\Gamma_j$.
6.4. Let F be a flag of $\bigoplus_{j\in J}\Gamma_j$ and $\Gamma_{j(F\cap S_j)}$ be the residue of $F \cap S_j$ in Γ_j. Then Γ_F is isomorphic to the direct sum of the geometries $\Gamma_{j(F\cap S_j)}$.
6.5. $\bigoplus_{j\in J}\Gamma_j$ is firm (resp.thick, thin) if and only if each Γ_j is firm (resp.thick, thin).
6.6. $\bigoplus_{j\in J}\Gamma_j$ is strongly connected if and only if each Γ_j is strongly connected.
6.7. $\mathrm{Aut}_\Delta(\bigoplus_{j\in J}\Gamma_j)$ (the group of all Δ-automorphisms) is isomorphic to the direct product $\prod_{j\in J}\mathrm{Aut}_{\Delta_j}(\Gamma_j)$. Furthermore the first group is

chamber-transitive if and only if each $Aut_{\Delta_j}(\Gamma_j)$ is chamber-transitive.

6.8. $\text{Rank} \bigoplus_{j\in J} \Gamma_j = \sum_{j\in J} \text{rank}\, \Gamma_j$.

We shall say that a direct sum is <u>non-trivial</u> if there is no empty component among the Γ_j and if there are at least two distinct components Γ_j.

As usual the usefulness of a direct sum concept is to give rise to conditions under which some given object decomposes into a direct sum of other (simpler) objects. We shall now report on a result of this kind which is the first non-trivial theorem in the theory and which plays a crucial role in all studies of diagram geometries.

7. <u>The basic diagram of a geometry</u>.

Let $\Gamma=(S,I,t)$ be a geometry over Δ.
We shall now introduce a graph structure on Δ, say $\Delta(\Gamma)$ induced by Γ, which we call the <u>basic diagram</u> of Γ because diagrams to be introduced later will appear as specializations of it.

A pair of distinct elements i,j of Δ are called <u>joined</u>, i.e. they constitute an edge of the basic diagram, whenever there is at least one flag F of cotype $\{i,j\}$ in Γ whose residue is not a generalized digon, i.e. whose residue is not a non-trivial direct sum of other geometries.

<u>Comments</u> 1) In [1] the basic diagram is used to introduce Theorem 2, up to the terminology.
Pasini [17] observed that the general theory developed in [3] does not require the full strength of the diagram concept as introduced in [1] but that it requires only the basic diagram. These ideas are made more explicit in the next sections.

7.1. <u>Proposition</u>. Let $\Gamma=(S,I,t)$ be a geometry over Δ and F a flag of cotype Δ' in Γ. Then the basie diagram $\Delta'(\Gamma_F)$ of the residue of F is a subgraph of the basic diagram $\Delta(\Gamma)$.

Proof. Straightforward.

<u>Comments</u> 2) In most well behaved geometries, $\Delta'(\Gamma_F)$ is actually an induced subgraph of $\Delta(\Gamma)$, i.e. two elements i,j of Δ' are joined in Δ' with respect to Γ_F if and only if they are joined in Δ with respect to Γ. Pasini [17] constructs examples in which $\Delta'(\Gamma_F)$ is not always

an induced subgraph of $\Delta(\Gamma)$.
Moreover he observed a mistake in the theory developed in [1], [3] based on these examples (see also further sections).

7.2. The fundamental lemma. (Tits [25], theorem 2 in [1]). Let Γ be a strongly connected geometry of finite rank over Δ. Let i,j be elements of Δ which are contained in distinct connected components of the basic diagram $\Delta(\Gamma)$. Then every i-element of Γ is incident with every j-element of Γ.

Proof. Clearly the rank of Γ is at least equal to 2. Proceed by induction on r=rank Γ. If $r=2$, Γ is a generalized digon and so the property holds. Hence we may assume $r \geq 3$. Let k be an element of $\Delta-\{i,j\}$ which may be assumed not to be in the connected component of i in $\Delta(\Gamma)$. Let V_i, V_j be i- and j-elements of Γ. In view of the strong connectedness there is a chain joining V_i and V_j in $\Gamma_i \cup \Gamma_j$ namely
$V_i \ I\ V_j^1 \ I\ V_i^1 \ I\ V_j^2 \ \ldots \ I\ V_j$. It suffices to show $V_i \ I\ V_j^2$ in order to end the proof by another induction on the length of the chain. In $\Gamma_{V_i^1}$, the elements V_j^1 and V_j^2 are joined by a chain of j- and k-elements in view of strong connectedness

$$V_j^1 \ I\ V_k \ I\ V_j^{11} \ I\ V_k^1 \ I\ V_j^{12} \ \ldots \ I\ V_j^2$$

The residue $\Gamma_{V_j^1}$ has a basic diagram on $\Delta-j$ which is a subgraph of the basic diagram $\Delta(\Gamma)$ by Proposition 7.1, hence i and k are in distinct connected components of it and induction applies. Therefore $V_i \ I\ V_k$. The same argument applies to Γ_{V_k} instead of $\Gamma_{V_j^1}$ to show that $V_i \ I\ V_j^{11}$ and so repeated use of this reasoning gives $V_i \ I\ V_j^2$.

7.3. Theorem. Let $\Gamma=(S,I,t)$ be a strongly connected geometry of finite rank over Δ. Let $(\Delta_j)_{j\in J}$ be the family of connected components of the basic diagram $\Delta(\Gamma)$. Let $\Gamma_j=(S_j,I_j,t_j)$ be the induced subgeometry of Γ over Δ_j, defined as follows:
$S_j=t^{-1}(\Delta_j)$, $I_j=I|S_j$, $t_j=t|S_j$.
Then Γ is isomorphic to the direct sum $\bigoplus_{j\in J} \Gamma_j$.

Proof. First of all, each Γ_j is indeed a geometry, i.e. it satisfies the transversality condition.
Secondly, Δ (resp. S) is the disjoint union of the Δ_j's (resp. S_j's). This takes care of conditions 1), 2), 4) in 6.1. As to condition 3) of 6.1 it is an immediate consequence of the fundamental lemma 7.2.

Comments. 3) Theorem 7.3 has been implicity used as a consequence of 7.2 by a number of authors, in particular Tits [25]. It is a valuable contribution of A.Valette [31] to have made it so explicit and clear as above, in a somewhat less general form however.

4) It is possible to say a little more than in theorem 7.3. First of all, a converse holds trivially, i.e. if Γ is a direct sum of geometries Γ_j over Δ_j, $j \in J$ then for $j \neq k$ in J, Δ_j and Δ_k are contained in distinct components of $\Delta(\Gamma)$.

Secondly, as a trivial consequence of the preceding remark, if $\Delta(\Gamma)$ is connected then Γ cannot be a non-trivial direct sum of other geometries. Therefore the direct summands of Γ in theorem 7.3 are the best possible, i.e. they cannot be decomposed in turn in smaller components. Moreover we get a "unique factorization of Γ in indecomposable factors".

5) Question. What happens for geometries of infinite rank ?

6) If G is a given graph it is not very difficult to convince onseself of the existence of geometries Γ whose basic diagram $\Delta(\Gamma)$ is isomorphic to G.

8. Shadows.

Most geometries needed in "real life" are obtained as sets of points equipped with distinguished subsets, i.e. their elements are identified with sets of points. The purpose of shadows is to develop this point of view for any geometry.

Let $\Gamma=(S,I,t)$ be a geometry over Δ and let $i \in \Delta$. For any flag F of Γ the i-shadow or shadow $\sigma_i(F)$ of F in Γ_i is the set of all elements of Γ_i incident with F. If we want to emphasize that we work inside Γ we write $\sigma_i(F,\Gamma)$ for $\sigma_i(F)$.

The i-space of Γ is the set Γ_i of all i-elements of Γ equipped with all possible i-shadows of flags and ordered by inclusion. It will be (abusively) denoted by Γ_i.

A condition which is extremely natural in view of our geometric traditions is the following.

(GL)(Linearity condition [29]). A geometry $\Gamma=(S,I,t)$ over Δ is called linear if for any $i \in \Delta$, the intersection of two shadows of flags in Γ_i is necessarily the shadow of a flag or the empty set.

Comments. 1) Instead of a condition on the intersection of two flags we may of course require a stronger version, namely on the intersection of any family of shadows.

2) A somewhat disturbing property is that the shadow of the empty flag is the entire set Γ_i instead of the empty set as one would expect.
3) All geometries of rank ≤ 1 have property (GL) trivially.
4) If $\Gamma = \bigoplus_{j \in J} \Gamma_j$ then it is straightforward to check that Γ is linear if and only if each Γ_j is linear.
5) Question. If Γ is linear, is each residue of Γ also linear ? This is quite unlikely. If so, another version of (GL) would require the same property to hold in every residue of Γ.
6) If Γ is linear then shadows of flags which "cover" points in the sense of lattice theory may be called lines. Then any two points are at most on one line and if a line intersects a shadow $\sigma(F)$ in two points then it is contained in $\sigma(F)$.

We shall now introduce a stronger version of the linearity condition which has turned out to be very useful in order to develop general theorems and which is observed quite often, though not always, in "real life".

(IP)(Intersection property)(axiom (3) in [1]). A geometry $\Gamma=(S,I,t)$ over Δ is said to have the intersection property if for each $i \in \Delta$, $x \in S$ and F a flag, then either $\sigma_i(x) \cap \sigma_i(F)$ is empty or there is a flag F' incident with x and F such that $\sigma_i(x) \cap \sigma_i(F)=\sigma_i(F')$ and moreover the same property holds in every residue of a flag in Γ.

Comments. 7) We notice that (IP) implies (GL) and that F' may be the empty flag in the statement of (IP).
8) Question. Does (GL) together with (GL) in every residue of Γ imply (IP) for Γ? This looks very unlikely but we have no available counter-example to submit.
9) There are geometries arising in a very interesting group theoretical context which do not satisfy (IP) nor (GL). This appeared already as a remark in [1]. It became fully apparent in the work of Ronan-Smith [22] and Kantor [12].

8.1. Proposition. Let $\Gamma = \bigoplus_{j \in J} \Gamma_j$ be a direct sum of geometries. Then Γ has the intersection property if and only if each Γ_j has the intersection property.

Proof. 1) Assume each Γ_j satisfies (IP) and let $i \in \Delta$, $x \in S$, F be a flag in Γ. Now $x \in \Gamma_k$ for some $k \in J$. We distinguish two cases.
1.1) $i \in \Delta_k$. Then x is an element of Γ_k, $F \cap S_k$ is a flag of Γ_k and by (IP) either $\sigma_i(x,\Gamma_k) \cap \sigma_i(F \cap S_k,\Gamma_k)=A$ is empty or there is a flag F' of Γ_k, incident to x and $F \cap S_k$ such that $\sigma_i(F',\Gamma_k)=A$.
Since $\sigma_i(x,\Gamma) \cap \sigma_i(F,\Gamma)=A$ in all cases we see that either this inter-

section is empty or that there is a flag of Γ, namely F' which is incident to x and to F and whose shadow is $\sigma_i(F',\Gamma)=A$.

1.2) $i \notin \Delta_k$. Let $i \in \Delta_\ell$, $\ell \in J$. Then $F'=F \cap S_\ell$ is incident with x and with F and $\sigma_i(F')=\sigma_i(x) \cap \sigma_i(F)$; hence (IP) holds in Γ.

2) Assume that Γ satisfies (IP). Then (IP) holds in every residue of Γ . As each Γ_j is clearly the residue of a flag of cotype Δ_j we get (IP) in Γ_j as well.

8.3. Proposition. Let Γ be a firm strongly connected geometry of finite rank over a set Δ in which the intersection property holds. Then

(i) every rank 2 residue of Γ is either a generalized digon or a partial linear space;

(ii) if $i \in \Delta$ **then any finite intersection of i-shadows of flags of Γ is** an i-shadow of a flag or is empty.

Proof.(i) Let Φ be a residue of rank 2 of Γ which is not a generalized digon. Then (IP) holds in Φ and we shall assume that Φ is a geometry over $\{0,1\}$. Let V,W be two 1-elements of Φ and let us assume that their o-shadows contain at least two elements a,b in common. Then we shall show that V=W and so that Φ is a partial linear space. By (IP) there is a flag X of Φ incident with V,W such that $\sigma_o(V) \cap \sigma_o(W)=\sigma_o(X)$. If X is non-empty, $\sigma_o(X)$ contains a,b and so X consists of a unique element which is of type 1. As XIV and XIW we get X=V=W.

If X is empty, then $\sigma_o(V)=\sigma_o(W)=\Phi_o$. As Φ is not a generalized digon there is some 1-element U such that $\sigma_o(U)\neq\Phi_o$. Then $\sigma_o(V) \cap \sigma_o(U)=\sigma_o(U)$ and there is a non-empty flag Y incident with V and U such that $\sigma_o(Y)=\sigma_o(U)$, in view of (IP). As Γ is firm, Φ is firm also and so $\sigma_o(U)$ contains at least two elements. Therefore Y must consist of a unique element which is a 1-element and so Y=U. Hence UIV and so U=V, a contradiction.

(ii) Let V,W be flags of Γ where W is a set of elements $W_1,\dots,W_n$ of Γ. We must show that if $\sigma_o(V) \cap \sigma_o(W)$ is non-empty then it is equal to the shadow of some flag. Now $\sigma_o(W)=\bigcap_{i=1}^{n}\sigma_o(W_i)$. On the other hand $\sigma_o(V) \cap \sigma_o(W_1)=\sigma_o(X_1)$ for some flag by (IP). Then $\sigma_o(V) \cap \sigma_o(W_1) \cap \sigma_o(W_2)=\sigma_o(X_1) \cap \sigma_o(W_2)=\sigma_o(X_2)$ for some flag X_2 again by (IP). Repeated use of this argument leads to the predicted conclusion.

9. Theory of pure geometries.

Let Γ be a strongly connected firm geometry of finite rank over Δ with the intersection property. Assume i,j are joined in the basic diagram $\Delta(\Gamma)$. Pasini [17] constructed examples of such geometries Γ in which there are nevertheless residues of type {i,j} which are generalized digons. Moreover he noticed that they contradict the theory as developed in [1], [3] and started repairing this to a large extent.

We shall say that a geometry Γ is pure (Pasini [17]) if for any vertices i,j of Δ which are joined in the basic diagram $\Delta(\Gamma)$ no residue of Γ of type {i,j} is a generalized digon. In other words, if Γ is pure then the basic diagram of Γ_F is the induced graph $\Delta(\Gamma)-t(F)$.

Under this additional assumption of purity the theory developed in [1], [3] holds together with the proofs given there. We shall outline this for the sake of clarity and develop some additional consequences.

From now on in section 9 all geometries are pure, firm, strongly connected, of finite rank, with the intersection property.

9.1. Control over separation in the basic diagram [2].

Consider the graph $\mathcal{G}=\Delta(\Gamma)$. Let o be a specified point of $\mathcal{G}$. If A,B are sets of points of $\mathcal{G}$, then A separates o from B or $A \preccurlyeq B$ if there is no path in $\mathcal{G}-A$ joining o to some point of B. Notice that o may belong to A, that separation $\preccurlyeq$ is transitive and reflexive and that it determines therefore a pre-order on the subsets of $\mathcal{G}$. If B',B" are subsets of B separating o from B, then $B' \cap B''$ separates o from B. Assume the contrary and let $o=X_o, X_1, \ldots, X_n=b$ be a path joining o to $b \in B$ which does not intersect $B' \cap B''$. Then some $x_i \in B'$ because B' separates o from B. Take the smallest i such that $x_i \in B'$. Suppose first that $i > 0$. As $x_i \notin B''$ and as B" separates o from B, there is some $j < i$ such that $x_j \in B''$. Now $X_j=o$, otherwise there would again be an element X_k, $k < j$ with X_k in B'. Hence we may assume $X_i=o$ or equivalently $X_j=o$. Take $X_i=o$, i.e. $o \in B'$. Now $o \notin B''$ and so B" cannot separate o from B since o is joined to itself by a path inside B. Hence we have proved that $B' \cap B''$ separates o from B. Therefore there is a smallest subset of B separating o from B which we call the o-reduction of B. The set of points B of $\mathcal{G}$ is called o-reduced if B is equal to its o-reduction. Notice that {o} is o-reduced. Let $R_o(\mathcal{G})$ denote the set of all o-reduced subsets of $\mathcal{G}$ provided with the separation relation $\leq$.

A set B of points of G will be o-connected if B contains o and if B is connected as an induced subgraph of G. Clearly the union of two o-connected sets is o-connected and so the set $C_o(G)$ of all o-connected subsets of G provided with the relation of inclusion is obviously a lattice.

We shall establish a close connection between o-connected and o-reduced sets and derive that $R_o(G)$ is a lattice.

Let A be some o-reduced subset of G other than {o}.
Then o $\notin$ A since otherwise {o} would be a subset of A separating o from A. The connected component A_o of o in G-A will be called the <u>interior</u> of A.
Let now C be an element of $C_o(G)$. Then the set $F_r(C)$ of all elements of G-C which are joined by an edge to some point of C is called the <u>frontier</u> of C.
Finally, let A be some o-reduced subset of G other than {o}. For any a $\in$ A the <u>derived set</u> A(a) is defined as (A-a) $\cup$ {x$\in$ G-(A$\cup A_o$)|x joined to a}.

9.2.<u>Lemma</u>. The mapping s: $R_o(G)$-{o}$\rightarrow$ $C_c(G)$ with A$\rightarrow A_c$ is a bijection; in particular C $\in$ $C_o(G)$ implies that Fr(C) $\in$ $R_o(G)$-{o}.

<u>Proof</u>.1) <u>s is onto</u>. Let C $\in$ $C_o(G)$ and consider Fr(C)=A. As o $\in$ C, A$\neq$ o and A is o-reduced since G-A contains C and so a proper subset of A could not separate o from A. Moreover, A_o contains C since C is connected and o $\in$ C. If a $\in$ A_o-C, then we may assume that a is joined to some point of C by an edge and so a $\in$ A: a contradiction. Therefore A_o=C, s($R_o(G)$- {o})= $C_o(G)$ and we see already that Fr(C) $\in$ $R_o(G)$-{o}.
2) <u>s is injective</u>. If C $\in$ $C_o(G)$ and B $\in$ $R_o(G)$ with B_o=C, then Fr(C) $\subseteq$ B, since otherwise B_o would contain C and the additional point p. Now A separates o from B: indeed let P be a path joining o to b $\in$ B and let c be the point of C $\cap$ P nearest b. Clearly c$\neq$b. Hence there is a unique point x on P, between c and b, joined to c. Now x $\in$ A, hence A separates o from B. Since B is o-reduced this implies A=B and so s is a bijection.
9.3.<u>Theorem</u>.(i) The mapping s is an isomorphism of ordered sets of $R_o(G)$-{o} onto $C_o(G)$.
(ii) $R_o(G)$ is a lattice all of whose maximal chains have the same cardinality (Jordan-Dedekind lattice).
(iii) if A is an o-reduced set of G then every subset of A is o-reduced as well.

(iv) if A is an o-reduced subset of G other than $\{o\}$ then an element X of $R_o(G)$ covers A (i.e. if $A \leq Y \leq X$ then $Y=A$ or $Y=X$) if and only if X is a derived set $A(p)$ where $p \in A$.

Proof. (i) In view of Lemma 9.2 it suffices to show that $A \leq B$ in $R_o(G)$ if and only if $A_o \subseteq B_o$. Assume first that $A_o \subset B_o$. Let P be a path from o to $b \in B$ and suppose by way of contradiction that P does not intersect A. Then $b \in A_o$. However $b \in B$ implies $b \notin B_o$ contradicting our hypothesis. Hence $A \leq B$. Next we assume $A \leq B$. Let $x \in A_o$ and let P be some path from o to x contained in A_o. If x is not in B_o, there is some point $b \in B \cap P$ separating o from x on P. As $A \leq B$ there is a point $a \in A$ on the subpath of P joining o to b and this contradicts $P \subset A_o$.
Therefore $x \in B_o$ and $A_o \subset B_o$.
(ii) follows from (i) and from the trivial observation that $C_o(G)$ is a Jordan-Dedekind lattice. Indeed, in $C_o(G)$ an element covers another one if and only if it has precisely one more element.
(iii) In view of Lemma 9.2 every element of A is joined by an edge to some element of A_o. If $B \subset A$ then it is clear that $B_o \supset A_o$ and so B is o-reduced as well since no proper subset can separate it from o.
(iv) First we consider some derived set $A(p)$ and we show that it covers A in $R_o(G)$. First of all $A(p)=Fr(A_o \cup \{p\})$ and so $A(p)$ is o-reduced by lemma 9.2. In addition to this, the isomorphism σ of ordered sets sends $A(p)$ on an element of $C_o(G)$ which covers A_o. Hence $A(p)$ covers A. Conversely, if X covers A then $\sigma(X)=X_o$ covers $\sigma(A)=A_o$ and so $X_o=A_o \cup \{p\}$ for some element p of G. As X_o is connected, lemma 9.2 shows that $p \in A$ and so $X=Fr(A_o \cup \{p\})=A(p)$.

Comment. 1) In view of theorem 9.3 (ii) a dimension can be defined on $R_o(G)$ in some standard way. Here we determine dimension by the requirement that $\{o\}$ has dimension o.

9.4. Definitions. We need to complete our definitions in order to state and prove our general theorems.
Let $\Gamma=(S,I,t)$ be a geometry over Δ and let o be a specified element of Δ. Then Δ_o is the connected component of o in the basic diagram $\Delta(\Gamma)$ and $\Delta(o)$ is the set of vertices of the basic diagram joined to o. A flag F of Γ is called o-reduced if no proper subflag F' of F has the same o-shadow, i.e. if $\sigma(F')\neq\sigma(F)$ for all proper subflags F' of F. Moreover, F is strongly o-reduced if there is no flag E with $|E| < |F|$ such that $\sigma_o(E)=\sigma_o(F)$. Clearly strong reduction implies reduction.

Finally o-shadows of elements of type o are called points and o-shadows of flags of type $\Delta(o)$ are called lines.

9.5. Lemma (theorem 4 of [1]). Let $\sigma_o(L)$ be a line (i.e. L is a $\Delta(o)$-element) and let V be an element of Γ such that $\sigma_o(L)$ intersects $\sigma_o(V)$ in at least two points. Then L I V.

Proof. Let p,q be distinct points in $\sigma_o(L) \cap \sigma_o(V)$. We proceed by induction on the rank r of Γ. If r=1, then $\Delta(o)$ is empty, L is empty and so L I V is obvious. Hence we assume $r \geqslant 2$ and L non-incident with V. By (IP) there is a flag X incident with L,V such that $\sigma_o(X)=\sigma_o(L) \cap \sigma_o(V)$. If X was empty then $\sigma_o(X)=\Gamma_o=\sigma_o(L)=\sigma_o(V)$. Let ℓ be an element of L (L is non-empty since L is not incident to V). Let Y be a chamber containing ℓ and let Y' be the subflag of Y of cotype $\{0,t(\ell)\}$. Then the residue $\Gamma_{y'}$ is of type $\{o,t(\ell)\}$ and it contains ℓ. In view of the purity of Γ, $\Gamma_{y'}$ cannot be a generalized digon. Therefore it is a partial linear space (Proposition 8.3,(i)). In the latter ℓ is incident with all points since $\sigma_o(L)=\Gamma_o$. Therefore ℓ is the only element of type $t(\ell)$ in $\Gamma_{y'}$ and this contradicts the firmness hypothesis made on Γ. Hence X is non empty and there is some element W in it which is therefore incident with p,q,L,V.

Now W cannot be an element of type o or of type t(V) since it is incident with p,q,L,V. We want to use some W such that $t(W) \notin \Delta(O)=t(L)$ Assume this does not occur. Then $t(X) \subset t(L)$ and as X I L this means that $X \subset L$. Now Γ_X is a geometry over $\Delta'=\Delta-t(X)$, $o \in \Delta'$, $\Delta'(o)=\Delta(o)-t(X)$ and $\sigma_o(L-X,\Gamma_X) \cap \sigma_o(V,\Gamma_X)$ contains p and q. Therefore induction applies and so (L-X)I V and finnaly L I V. If on the other hand we have some W with $t(W)\notin\Delta(o)$ then Γ_W is a geometry over $\Delta''=\Delta-t(W), o \in \Delta''$, $\Delta''(o)=\Delta(o)$ and $\sigma_o(L,\Gamma_W) \cap \sigma_o(V,\Gamma_W)$ contains p,q; hence by the induction hypothesis L I V.

9.6. Lemma (Lemma 1 in [3])
If V is an element of Γ such that $\sigma_o(V)=\Gamma_o$ then t(V) does not belong to Δ_o, i.e. the empty set separates o from t(V) in Δ.

Proof. We proceed by induction on the rank r of Γ. Clearly $r \geqslant 2$. For r=2, Γ must be a generalized digon and $t(V)\neq o$ suffices to get the required property; this results from Proposition 8.3 (i) and the fact that a partial linear space with a line $\sigma_o(V)$ containing all points can have but one line and this contradicts firmness. Hence we may assume $r \geqslant 3$. Then there is some element W of Γ, with W I V,

$o \neq t(W) \neq t(V)$. In Γ_W we find V and we still have $\sigma_o(V,\Gamma_W)=(\Gamma_W)_o$, hence induction applies and t(V) is not in $(\Delta-t(W))_o$, i.e. t(W) separates o from t(V). Assume by way of contradiction that there is a path P from o to t(V) in $\Delta(\Gamma)$. Then the last argument shows that P must involve every element of Δ, i.e. $\Delta(\Gamma)$ is a linear graph with o and t(V) as endpoints. Here $\Delta(o)$ is reduced to a single element. Let L be any $\Delta(o)$-element. Then $\sigma_o(L)$ intersects $\sigma_o(V)$ in two points at least since Γ is firm and $\sigma_o(V)=\Gamma_o$. By lemma 9.5 we get L I V. Therefore, if p is any o-element, then we get $\sigma_{\Delta(o)}(V,\Gamma_p)=(\Gamma_p)_{\Delta(o)}$ and so the induction hypothesis shows that t(V) does not belong to $(\Delta-o)_{\Delta(o)}$ which is contradictory with the fact that $\Delta(\Gamma)$ is a linear graph. The conclusion is that there was no path P from o to t(V) in $\Delta(\Gamma)$.

9.7.Theorem (theorem 1 in [31])

Let A and B be o-reduced flags of Γ. Then $\sigma_o(A) \subset \sigma_o(B)$ if and only if $A \cup B$ is a flag and t(A) separates o from t(B) in Δ. Moreover $\sigma_o(A)= \sigma_o(B)$ if and only if A=B.

Proof. In view of Lemma 7.2, since A and B are o-reduced, t(A) and t(B) are contained in Δ_o. If $A \cup B$ is a flag and t(A) separates o from t(B) then Lemma 7.2 applied to Γ_A shows that $\sigma_o(B,\Gamma_A)=(\Gamma_A)_o$ and so $\sigma_o(A) \subset \sigma_o(B)$. We shall now prove the converse. Therefore we assume $\sigma_o(A) \subset \sigma_o(B)$. If A I B then we apply Lemma 9.6 to Γ_A and to every element V of B not in A getting that t(V) does not belong to $(\Delta-t(A))_o$; hence t(A) separates o from t(B).

Hence we have only to show that A I B. We may assume without loss of generality that B is reduced to a single element of Γ. We proceed by induction on the rank r of Γ. For r=1 the property is trivial. Hence we suppose $r \geqslant 2$. If some element A_i of A is of type o then $A= A_i$ because A is o-reduced and so A I B. Hence we assume that $o \notin t(A)$ and we distinguish two cases.

Case 1. Some element A_i of A is incident with B. In the residue Γ_{A_i} we find $A-A_i$, B and $\sigma_o(A-A_i,\Gamma_{A_i}) \subset \sigma_o(B,\Gamma_{A_i})$. Obviously $A-A_i$ is o-reduced in Γ_{A_i}. If B is also o-reduced in Γ_{A_i} then induction applies and we get A I B. If B is not o-reduced in Γ_{A_i} then $\sigma_o(B,\Gamma_{A_i})=(\Gamma_{A_i})_o$ and by lemma 9.6. the empty set separates o from t(B) in $\Delta-t(A_i)$; hence $t(A_i)$ separates o from t(B) in Δ. As A is o-reduced, $t(A_i)$ does not separate o from $t(A_j)$ for any $A_j \in A-A_i$ (lemma 7.2) and so $t(A_i)$ separates $t(A_j)$ from t(B) since otherwise there would be a path from o to t(B) avoiding $t(A_i)$. Therefore, in Γ_{A_i} lemma 7.2 shows that

A_j I B and finally AIB.

Case 2. No element of A is incident with B. Then we derive a contradiction. By (IP) there is a flag X incident with A and B such that $\sigma_o(X)=\sigma_o(A)\cap\sigma_o(B)=\sigma_o(A)$. Now X is non-empty since otherwise $\sigma_o(X)=\sigma_o(A)=\sigma_o(B)=\Gamma_o$ and then lemma 9.6 shows that B is not o-reduced. Let X' be the o-reduction of X. Then $\sigma_o(A)=\sigma_o(X')$ and case 1 show that t(A) separates o from t(X') and that t(X') separates o from t(A). Hence t(A)=t(X'). Indeed, if there is for instance an element $a \in t(A)-(t(A)\cap t(X'))$ let P be a path in Δ from o to a. Then P contains a member x of t(X') and P contains a member a' of t(A) between o and x. As $a \notin t(X')$, $x \neq a$ and so $a' \neq a$. Therefore t(A)-a separates o from a and then Lemma 7.2 applied to Γ_{A-a} shows that A is not o-reduced.

As t(A)=t(X') and AIX' we see that A=X' and so we get AIB as we needed.

As a final step we have to study the case $\sigma_o(A)=\sigma_o(B)$. From the first part of the theorem we get AIB and that t(A) separates o from t(B) and t(B) separates o from t(A). This is the situation we analyzed earlier (see A and X') and so t(A)=t(B) and finally A=B.

9.8.Corollary. Any o-reduced flag is strongly o-reduced.

9.9.Theorem (theorem 2 in [31]).
Let X be a subset of Δ. Then a flag of type X is o-reduced if and only if X is o-reduced in Δ.

Proof.1) Assume X is not o-reduced. Then there is a proper subset X' of X separating o from X. Let F be some flag of type X and let F' be the subflag of F such that t(F')=X'. Then Lemma 7.2 applied to $\Gamma_{F'}$ shows that $\sigma_o(F')=\sigma_o(F)$ and so F is not o-reduced.
2) Assume F is a flag of type X and F' is a proper subset of F such that $\sigma_o(F')=\sigma_o(F)$. Then Lemma 9.6 applied to $\Gamma_{F'}$ shows that t(X)=X' separates o from X and so X is not o-reduced.

9.10.Theorem (theorem 3 in [31]).
Let $\Delta(\Gamma)$ be connected and let V,W be elements of Γ. Assume that A is a o-reduced subset of Δ separating pairwisely o,t(V), t(W). Then VIW if and only if there is some flag F of type A such that $\sigma_o(F)\dot{\in}\sigma_o(V)\cap\sigma_o(W)$.

Proof. 1) Assume first that $\sigma_o(V)\cap\sigma_o(W)\supset\sigma_o(F)$ where F is some flag of type A. We may assume F o-reduced of course. Then theorem 9.7 shows that FIV and FIW. If either $V\in F$ or $W\in F$ we get VIW as we want. Otherwise we see that t(V), t(W) are in distinct connected components of Δ-A and Lemma 7.2 shows that VIW in Γ_F.

2) If VIW then the flag {V,W} may be completed by some flag of type A, say X, and in Γ_X the empty set separates o from t(V). Hence by lemma 7.2, if $t(V)\notin A$, then $\sigma_o(V,\Gamma_X)\supset(\Gamma_X)_o$ and similarly for W; therefore $\sigma_o(V)\cap\sigma_o(W)\supset\sigma_o(X)$ whenever $t(V)\notin A$ and $t(W)\notin A$. If $t(V)\in A$ then $V\in X$ and $\sigma_o(V)\supset\sigma_o(X)$ anyway; hence the conclusion is $\sigma_o(V)\cap\sigma_o(W)\supset\sigma_o(X)$ in all cases.

9.11. Theorem (theorem 4 in [3]). Assume that $\Delta(o)$ is non-empty in Δ.
(i) If p,q are distinct o-elements of Γ then there is at most one flag of type $\Delta(o)$ incident with p and q.
(ii) If V is some element of Γ with $t(V)\in\Delta_o$ and if L is some flag of type $\Delta(o)$ such that $\sigma_o(L)\cap\sigma_o(V)$ contains at least two points, then $L\cup V$ is a flag and $\sigma_o(L)\subset\sigma_o(V)$.

Proof. (i) Let L_1,L_2 be flags of type $\Delta(o)$ incident with p and q. By (IP) there is a flag X incident with L_1,L_2 such that $\sigma_o(X)=\sigma_o(L_1)\cap\sigma_o(L_2)$ and we may assume that X is o-reduced. Since $\Delta(o)$ is obviously o-reduced theorem 9.7 applies to $\sigma_o(X)\subset\sigma_o(L_1)$ and so t(X) separates o from $t(L_1)=\Delta(o)$. This means that either $o\in t(X)$ or that $\Delta(o)\subset t(X)$. In the first case, $\sigma_o(X)$ is reduced to a single point while p,q are distinct members of it: a contradiction. Hence $\Delta(o)\subset t(X)$ and as L_1IX we get $L_1\subset X$. Similarly $L_2\subset X$ and so $L_1=L_2$ by the transversality property.
(ii) If LIV then we see in Γ_L that $\sigma_o(L)\subset\sigma_o(V)$ in view of Lemma 7.2. Hence it suffices to show LIV. Consider a member L_a of L. Then $\sigma_o(L_a)\cap\sigma_o(V)\supset\sigma_o(L)\cap\sigma_o(V)$ is non-empty and by (IP) there is an o-reduced flag X incident to L_a and to V such that $\sigma_o(X)=\sigma_o(L_a)\cap\sigma_o(V)$. Then $\sigma_o(X)\subset\sigma_o(L_a)$ and as L_a is also o-reduced, theorem 9.7 shows that t(X) separates o from $t(L_a)$. However, o and $t(L_a)$ are joined in $\Delta(\Gamma)$, hence $o\in t(X)$ or $t(L_a)\in t(X)$. If $o\in t(X)$, $\sigma_o(X)$ contains a unique point in contradiction with the assumptions. Therefore $t(L_a)\in t(X)$, $L_a\in X$ and finally L_aIV as we had to show.

Consider the space Γ_o i.e. the set of all o-elements of Γ equipped with all o-shadows of flags and ordered by inclusion. For any shadow σ in Γ_o there is a unique o-reduced flag F of Γ such that $\sigma_o(F)=\sigma$

(theorem 9.7). Hence it makes sense to define the mapping

$\tau: \Gamma_o \to \mathcal{R}_o(\Delta(\Gamma))$ with $\sigma \to t(F)$

which assigns to every shadow in the space Γ_o the type in Δ of the o-reduced flag which determines the shadow.

9.12. Theorem. The mapping τ is a surjective isometry, i.e. it preserves order and dimension or equivalently σ_1 covers σ_2 in Γ_o implies that $\tau(\sigma_1)$ covers $\tau(\sigma_2)$ in $\mathcal{R}_o(\Delta(\Gamma))$.
In particular, maximal sets of o-shadows which are totally ordered by inclusion have cardinality $|\Delta_o|+1$.

Proof. 1) Clearly τ is surjective.
2) Assume that σ_1 covers σ_2 in Γ_o. Then there are o-reduced flags A,B of Γ such that $\sigma_o(A)=\sigma_1$ and $\sigma_o(B)=\sigma_2$. By theorem 9.7 t(B) separates o from t(A) in $\Delta(\Gamma)$, i.e. $\tau(\sigma_1) \geq \tau(\sigma_2)$.
Now $\tau(\sigma_1) \neq \tau(\sigma_2)$ because AIB (theorem 9.7) and otherwise A=B which implies $\sigma_1=\sigma_2$.
Also if $\tau(\sigma_1) \geq T \geq \tau(\sigma_2)$ in $\mathcal{R}_o(\Delta)$ then there exists a flag X of type T which is incident with A and B (for instance in some chamber containing A and B). Now X is o-reduced (theorem 9.9). By theorem 9.7 we get $\sigma_1=\sigma_o(A) \supset \sigma_o(X) \supset \sigma_o(B)=\sigma_2$ and as σ_1 covers σ_2, $\sigma_o(X)=\sigma_1$ or $\sigma_o(X)=\sigma_2$; hence X=A or X=B (theorem 9.7) and so $\tau(\sigma_1)$ covers $\tau(\sigma_2)$.
This proves our first statement. Consequently τ maps a maximal set of o-shadows which is totally ordered by inclusion onto a maximal chain of $\mathcal{R}_o(\Delta(\Gamma))$ and the latter have cardinality $|\Delta_o|+1$ in view of Theorem 9.3.

The set of all o-elements of Γ is of course a o-shadow, namely the shadow of the empty flag. As usual, the term <u>maximal o-shadow</u> will denote a maximal proper element in the lattice of all o-shadows, i.e. a hyperplane.

9.13. Theorem. Every o-shadow of Γ is the intersection of a family of maximal o-shadows.

Proof. Let F be an o-reduced flag. If F is empty then an empty family of maximal o-shadows satisfies our requirement. Hence we assume F to be non-empty. If F consists of elements $F_1,\ldots,F_n$ of Γ then $\sigma_o(F)=\bigcap_{i=1}^{n}\sigma_o(F_i)$ and so it suffices to show that each $\sigma_o(F_i)$ is the intersection of a family of maximal o-shadows. Therefore we assume that F is an element of Γ and so t(F)=a is some element of Δ. We may assume that $\sigma_o(F)$ is not itself a maximal shadow. Let $p \in \Gamma_o - \sigma_o(F)$. It súffices to show that there is some maximal o-shadow containing $\sigma_o(F)$ which does not contain p.

Therefore it suffices to show that there are at least two o-shadows covering $\sigma_o(F)$ in the space Γ_o since then their intersection is necessarily $\sigma_o(F)$ and so one of them at least does not contain p which allows to produce inductively a maximal o-shadow containing $\sigma_o(F)$ and not p.

If A={a} consider the derived set A(a). By theorem 9.12 a flag X whose o-shadow covers $\sigma_o(F)$ has a type covering A in $R_o(\Delta(\Gamma))$ and by Theorem 9.3 (iv) this means that X is of type A(a).

Therefore A(a) is non empty and $a \notin A(a)$.

In Γ_F there are at least two flags X_1, X_2 of type A(a) in view of the transversality condition (section 2). By theorem 9.7, $\sigma_o(X_1) \cap \sigma_o(X_2)$ contains $\sigma_o(F)$ and by theorem 9.12 (second statement) $\sigma_o(X_1)$ and $\sigma_o(X_2)$ cover $\sigma_o(F)$.

9.14. <u>Theorem</u>. Let F and G be o-reduced flags such that $\sigma_o(F) \cap \sigma_o(G)$ is non-empty. Then there exists a flag X incident to F and to G such that $\sigma_o(X) = \sigma_o(F) \cap \sigma_o(G)$.

<u>Proof</u>. Let $F = \{F_1, \dots, F_n\}$ where F_i is an element of Γ. Then $\sigma_o(F_1) \cap \sigma_o(G) = \sigma_o(X_1)$ where X_1 is some o-reduced flag incident with F_1 and G, in view of (IP). Now $\sigma_o(F_2) \cap \sigma_o(F_1) \cap \sigma_o(G) = \sigma_o(F_2) \cap \sigma_o(X_1) = \sigma_o(X_2)$ where X_2 is some o-reduced flag incident with F_2 and X_1 in view of (IP). As $\sigma_o(X_2) \subset \sigma_o(G)$ we see that $X_2 I G$ (theorem 9.7) and similarly $X_2 I F_1$. Proceeding inductively along the same line of reasoning we get eventually some o-reduced flag X_n such that $\sigma_o(X_n) = \sigma_o(F) \cap \sigma_o(G)$ and such that $X_n I F$, $X_n I G$.

9.15. <u>Theorem</u>. Assume V,W,W' are elements of Γ such that VIW, VIW', $t(W) = t(W')$ and $\sigma_o(V) \cap \sigma_o(W) = \sigma_o(V) \cap \sigma_o(W')$. Then either W=W' or $\sigma_o(V) \subset \sigma_o(W) \cap \sigma_o(W')$.

<u>Proof</u>. We shall assume that $\sigma_o(V)$ is not contained in $\sigma_o(W) \cap \sigma_o(W')$ and we may even assume that $\sigma_o(V)$ is not contained in $\sigma_o(W)$. If $\sigma_o(W) \subset \sigma_o(V)$ then $\sigma_o(W) \subset \sigma_o(W')$, hence by theorem 9.7, WIW' and W=W'. If $\sigma_o(W)$ is not contained in $\sigma_o(V)$ then {V,W} is a o-reduced flag. By theorem 9.9 {V,W'} is also o-reduced and so theorem 9.7 applied to Γ_V shows that $\sigma_o(W, \Gamma_V) = \sigma_o(W', \Gamma_V)$ and that WIW' hence W=W'.

9.16. Theorem. Let V,W be elements of Γ. Then VIW if and only if $\sigma_o(V) \cap \sigma_o(W)$ is maximal among all sets $\sigma_o(X) \cap \sigma_o(Y)$ where X and Y are elements of Γ such that $t(X)=t(V)$ and $t(Y)=t(W)$.

Proof. 1) Assume VIW and let us prove that $\sigma_o(V) \cap \sigma_o(W)$ is maximal. If $\sigma_o(X) \cap \sigma_o(Y) \supseteq \sigma_o(V) \cap \sigma_o(W)=\sigma_o(V,W)$ and $t(X)=t(V)$, $t(Y)=t(W)$ then theorem 9.7 shows that $XI\{V,W\}$ and $YI\{V,W\}$ provided that $\{V,W\}$ is o-reduced. Then we get immediately $X=V$ and $Y=W$. If $\{V,W\}$ is not o-reduced we may assume that $\sigma_o(V) \subset \sigma_o(W)$ which implies XIV, YIV and $X=V$ again. In Γ_V, $\sigma_o(W,\Gamma_V) \supseteq \sigma_o(Y,\Gamma_V)$; hence WIY and therefore $W=Y$.

2) Assume that $\sigma_o(V) \cap \sigma_o(W)$ is maximal. Then it is non-empty and there is an o-reduced flag X such that $\sigma_o(X) = \sigma_o(V) \cap \sigma_o(W)$ with XIV, XIW, in view of (IP). By theorem 9.7, $t(X)$ separates o from $t(V)$. Let V' be an element of type $t(V)$ which is incident to $X \cup W$. Then theorem 9.7 implies $\sigma_o(X) \subseteq \sigma_o(V')$ and so $\sigma_o(V') \cap \sigma_o(W) \supseteq \sigma_o(V) \cap \sigma_o(W)$ and as the latter is maximal we get $\sigma_o(V') \cap \sigma_o(W)=\sigma_o(V) \cap \sigma_o(W)$. Now theorem 9.15 applies and so either $V=V'$ or $\sigma_o(W) \supseteq \sigma_o(V) \cap \sigma_o(V') \subseteq \sigma_o(V)$. In both cases we have WIV as required.

Comments. 2) An important underlying motivation of the preceding results is to reconstruct Γ and Δ from the knowledge of the space Γ_o i.e. of the set of o-elements together with all o-shadows of flags of Γ. Actually such a program is too ambitious since there are non isomorphic geometries having isomorphic spaces Γ_o. An example is provided by an hyperbolic quadric which arises as a geometry over the basic diagram

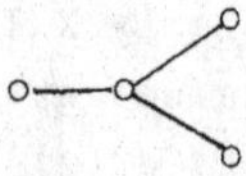

and as a geometry over .

A somewhat less ambitious question is however completely answered by our results as we shall see now.

3) Assume the space Γ_o is given as well as the partitioning of shadows in types. Then Γ can be reconstructed from this information. Therefore one has to be able to decide which shadows or types correspond to single elements of Γ or Δ.

Assume σ is a shadow. If it is maximal then it corresponds clearly to a single element of Γ. If it is not maximal then two cases are possible:

a) $\sigma=\sigma_o(F)$ where F is an element of Γ. Then all shadows which cover σ are of the same type.

b) $\sigma=\sigma_o(F)$ where F is o-reduced and F contains more than one element of Γ. Then all shadows which cover σ are of distinct types. These statements are consequences of 9.12 and 9.3 (iv). Hence we can reconstruct the set S of all elements of Γ , the set Δ and the mapping t. Finally I is reconstructed thanks to 9.16.

4) There is some hope that more can be said. If only the space Γ_o is given then Γ cannot be uniquely reconstructed. However among all geometries giving rise to the same space Γ_o there might be a kind of canonical one as is suggested by the example of quadrics mentioned in 2).

10. Pasini's characterization of pure geometries.

We shall always assume here that Γ is a strongly connected firm geometry of finite rank over a set Δ and that Γ satisfies (IP).

Pasini has developed a remarkable theory of these geometries showing among many other things that they are pure whenever the basic diagram has no 3-cycles. His first result is as follows.

10.1 Theorem (Pasini [17]). For each type $i \in \Delta$ and each choice of non empty i-reduced flags A,B in Γ if AIB and $\sigma_i(A)=\sigma_i(B)$ then A=B. The proof of this result is by no means trivial. It has consequences in the direction of a generalized theorem 9.7 which are not quite completely clear to the author.

10.2 Theorem [17] Γ is pure if and only if the following condition is satisfied for every choice of distinct types i,j in Δ and for every choice of flags F,F' of cotype {i,j}: if $\sigma_i(F) \cap \sigma_i(F')$ and $\sigma_j(F) \cap \sigma_j(F')$ have each at most one element then Γ_F and $\Gamma_{F'}$ are both generalized digons or they are both partial planes.

Pasini gives a nice condition on the basic diagram which implies purity of the geometry.

10.3 Theorem [17]. If $\Delta(\Gamma)$ has no cycles of length three then Γ is pure.

11. Specializations of diagrams.

A special diagram (Δ,f) on a set Δ is a function f which assigns to every ordered pairs of distincts elements (i,j) of Δ some class $\Delta_{ij}=f(i,j)$ of rank 2 geometries over {i,j} such that

(1) either all members of Δ_{ij} are generalized digons or all members of Δ_{ij} are partial planes

(2) $\Delta_{ji}=\Delta^*_{ij}$ where Δ^*_{ij} is the dual class of Δ_{ij} and the dual of a rank

2 geometry $\Gamma=(S,I,t)$ over $\{i,j\}$ is the geometry $\Gamma^*=(S,I,t^*)$ where t^* is defined by $t^*(V)\neq t(V)$ for all $V \in S$.

(3) Δ_{ij} is closed under isomorphisms.

A geometry Γ over Δ belongs to the diagram (Δ,f) if for every ordered pair of distinct elements i,j in Δ and every flag F of Γ of cotype $\{i,j\}$, the residue Γ_F is a member of Δ_{ij}.

Comments 1) The definition of special diagram given here is somewhat restricted with respect to the diagram definition given in [11]. This is harmless as far as applications are known to the author.

2) The restriction (1) in the definition of a special diagram refers clearly to the purity condition introduced earlier.

3) The basic diagram of a pure geometry Γ can be seen as a special diagram actually as the universal special diagram to which Γ belongs. If i,j are joined in $\Delta(\Gamma)$ then Δ_{ij} is the class of all partial linear spaces and if i,j are not joined then Δ_{ij} is the class of all generalized digons. Hence a special diagram to which Γ belongs appears as a restriction or a specialization of the basic diagram.

We shall not formalize these concepts since there is no theoretical need for it so far.

12. Coxeter diagrams.

Coxeter diagrams is a class of special diagrams which has been widely studied. In this case there is an integer $m_{ij} \geq 2$ attached to every pair of distinct elements i,j of Δ and Δ_{ij} is the class of all generalized m_{ij}-gons. A Coxeter diagram (Δ,f) being given Tits (unpublished) has shown the existence of thick geometries belonging to it provided there are no subdiagrams of type

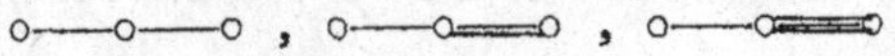

with the usual conventions on , ,  (generalized 3-, 4-, 5- gons). The method of his construction is a generalization of the construction of free projective planes initiated by M.Hall [10]. The main purpose of Tits [30] is to show that universal covers of geometries belonging to Coxeter diagrams are essentially buildings. In addition to this, Tits (unpublished) uses the amalgamation methods initiated for projective planes by Kegel-Schleiermacher [13] in order to construct "free buildings" with a BN-pair, i.e. with a "large" automorphism group, over any Coxeter diagram avoiding the subdiagrams mentioned earlier.

Comment 4) The moral of Tits' constructions is that geometries over a Coxeter diagram should not be considered as tight even under strong transitivity assumptions on the automorphism group corresponding to the existence of BN-pairs. Of course there are remarkable counter-examples to this statement. The geometries over a diagram of rank ⩾ 3 as

o——o——o ... o——o

o——o——o ... o══o

are indeed tight and they can be classified (projective spaces, polar spaces); those over o——o≡≡≡o are even more tight since there are no thick geometries over this diagram (Tits [28]).

13. Diagrams for sporadic groups.

The geometric analysis of sporadic groups carried out by the author (see [1], [4]) has shown that generalized n-gons do not suffice in this respect. As stated in [5] there is some evidence that a slight generalization could suffice. Therefore we introduced a generalized (g,d_p,d)-gon $(2 \leq g \leq d_p \leq d)$ as a firm rank 2 geometry in which the smallest circuits have 2g elements, in which each point (or o-element) is at greatest distance d_p from some other element, in which each line (or 1-element) is at greatest distance d from some other element and $d_p \leq g+2$.

A generalized Coxeter diagram is a special diagram (Δ,f) such that for all distinct i,j in Δ, Δ_{ij} is a class of generalized (g,d_p,d)-gons where g,d_p,d depend only on i and j.
For each of the 26 sporadic simple groups we have obtained at least one geometry which is strongly connected, pure, firm, satisfies (IP) which belongs to some generalized Coxeter diagram. We shall discuss this subject elsewhere.

For other work on diagram geometries related to sporadic groups see [15], [12], [22].

14. Linear diagrams.

For a characterization of geometries (pure, firm, etc) belonging to a linear diagram o $\overset{\Pi}{—}$ o $\overset{\Pi}{—}$ o ... o $\overset{\Pi}{—}$ o (resp. o $\overset{L}{—}$ o $\overset{L}{—}$ o ... o $\overset{L}{—}$ o) where o $\overset{\Pi}{—}$ o (resp. o $\overset{L}{—}$ o) stands for the class of all partial linear spaces (resp.all linear spaces) see [1].
The case of o $\overset{\rfloor}{—}$ o $\overset{L}{—}$ o has been masterfully classified by A.Sprague [23] and this result is likely to have consequences in various

directions.
The case of $o \overset{\subset}{\text{---}} o \overset{\supset}{\text{---}} o$ has been studied by D.Hughes and a number of other people (see paper of Hughes in these Proceedings). For other related work see Kantor [11], Percsy [19].

15. Apartments.

The analysis of early geometries related to sporadic groups made apparent the existence of certain subgeometries looking very much like the apartments of a building geometry.

In the theory of buildings apartments provide the major axiom: any two chambers are contained in some apartment. This does not hold any more in the sporadic geometries. On the other hand there is Tits' [30] characterization of buildings without any use of the apartment concept. For all of these reasons apartments have appeared somewhat out of the main stream and they have been given little attention. Still they are there ! What can be said about apartments of a geometry Γ ? First of all the concept is certainly not understood completely; i.e. we do not know what the right definition is or better what a useful definition is, one which would be verified on the most important geometries and which would allow more classification work.
An apartment of a geometry Γ must certainly be a thin (induced ?) subgeometry of Γ, having the same basic diagram and maybe the same generalized Coxeter diagram as Γ if Γ belongs to one.
Very often but not always, an apartment arises as the set of fixed elements of Γ by a specified subgroup of the automorphism group namely a subgroup H obtained as a complement in B of a nilpotent normal subgroup U, where B is the stabilizer of some chamber.

It would be interesting to further investigate this concept and to analyze how apartments and chambers interact in known geometries.

The author's work on the analysis of apartments is unpublished. See Kantor [12] for some examples.

16. Characterizing diagram geometries in terms of points and lines.

This important and vast subject with an enormous potential of applications in and outside geometry would deserve a long survey by itself. Here we shall only mention some of the most recent and most remarkable achievements. In the diagrams mentioned below the vertex corresponding to points is circled.
Cooperstein [7] has characterized the finite geometries of type

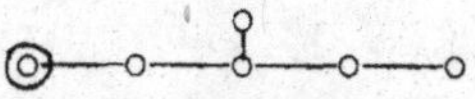

○─○─○̇─○─○ and he has done actually a lot more in several directions.

A.Cohen [6] has found a new system of axioms for the geometries of type ◎─○━○─○ (see also [27]) in a way which looks promising for future progress.

G.Tallini [24] has characterized ○─◎─○ ... ○─○ and this might be another starting point of useful developments.

Final comments made after the lectures

1. Questions 5) in section 7, 5) in section 8 and 8) in section 8 have been answered negatively by A. Brouwer who constructed a counter-example in each case. N. Percsy also produced a counter-example for the first of these questions.

2. Comment 6) in section 8 is wrong as pointed out by A. Brouwer.

17. References.

[1] Buekenhout,F. Diagrams for geometries and groups. J.Comb.Th.(A) 27 (1979), 121-151.

[2] Buekenhout, F.Separation and dimension in a graph. Geo.Ded. 8 (1979), 291-298.

[3] Buekenhout,F. The geometry of diagrams. Geo.Ded. 8 (1979),253-257

[4] Buekenhout,F. The geometry of diagrams. Proc.Symp.Pure Math. vol.34, AMS (1979), 69-75.

[5] Buekenhout,F. (g,d^*,d)-gons. Proceedings conference in honor of T.Ostrom. Pullman 1981 (to appear)

[6] Cohen,A. On the points and lines of metasymplectic spaces. (to appear)

[7] Cooperstein,B. A characterization of some Lie incidence structures. Geo.Ded. 6 (1977), 205-258.

[8] Fischer,B. Finite groups generated by 3-transpositions. Invent. Math. 13 (1971), 232-246.

[9] Guillite,S. and Percsy-Lefèvre,Ch. Pliages de géométries minces. Geo.Ded. (to appear)

[10] Hall,M. Projective planes. Trans.Amer.Math.Soc. 54 (1943),229-277.

[11] Kantor,W.M. Locally polar lattices. Journ.Comb.Th.(A) 26 (1979) 90-95.

[12] Kantor,W.M. Some geometries that are almost buildings. (to appear)

[13] Kegel,O. and Schleiermacher, A. Amalgams and embeddings of projective planes. Geo.Ded. 2 (1973), 379-395.

[14] Moore, E.H. Tactical Memoranda. Amer.J.Math. 18 (1896) 264-303.

[15] Neumaier,A. Rectagraphs, diagrams and Suzuki's sporadic simple group. (to appear).

[16] Ott,U. Bericht über Hecke Algebren und Coxeter Algebren endlicher Geometrien. Finite Geometries and Designs. Ed.by Cameron, Hirschfeld, Hughes. Cambridge U.P. 1981 p.260-271.

[17] Pasini,A. Diagrams and incidence structures. Preprint. Rapporto matematico n.21, Istituto di Mat, Univ.Siena, 1980

[18] Percsy,N. Plongement de géométries. Thesis Univ.Bruxelles, 1980.

[19] Percsy,N. Embedding geometric lattices in a projective space. "Finite geometries and designs" Ed.by Cameron, Hirschfeld, Hughes. Cambridge U.P. 1981, p.304-315.

[20] Ronan,M. Coverings and automorphisms of chamber systems. Europ.J.Comb. 1 (1980), 259-269.

[21] Ronan,M. Coverings of certain finite geometries. Finite geometries and designs. Ed. by Cameron, Hirschfeld, Hughes. Cambridge U.P. 1981, p.316-331.

[22] Ronan,M. and Smith,S. 2-local geometries for some sporadic groups. Proc.Symp.Pure Math. vol.37. AM.S. 1981

[23] Sprague,A. Rank 3 incidence structures admitting dual-linear, linear diagram. (to appear).

[24] Tallini,G. On a characterization of the Grassman manifold representing the lines in a projective space "Finite Geometries and Designs". Ed.by Cameron, Hirschfeld, Hughes. Cambridge U.P. (1981), 354-358.

[25] Tits,J. Les groupes de Lie exceptionnels et leur interprétation géométrique. Bull.Soc.Math.Belg. 8 (1956), 48-81.

[26] Tits,J. Sur la trialité et certains groupes qui s'en déduisent. Publ.Math. I.H.E.S. 2 (1959), 14-60.

[27] Tits,J. Buildings of spherical type and finite BN-pairs. Lecture Notes in Math, 386, Springer-Verlag, Berlin, 1974.

[28] Tits,J. Classification of buildings of spherical type and Moufang polygons: a Survey.
Atti Convegni Lincei 17 Roma (1976), 229-246.

[29] Tits,J. Buildings and Buekenhout Geometries. "Finite Simple Groups II" ed.M.Collins, Academic Press, New York, 1981,309-320

[30] Tits,J. A local characterization of buildings. Collection of papers dedicated to H.S.M. Coxeter (to appear).

[31] Valette,A. Direct sums of Tits geometries. Simon Stevin (to appear)

LINEAR SPACES AND STEINER SYSTEMS

Jean DOYEN

Department of Mathematics
Campus Plaine C.P.216
University of Brussels
B-1050 Brussels, BELGIUM

1. Introduction.

A linear space is a non-empty set S whose elements are called points, provided with a family of distinguished subsets called lines such that any two distinct points are contained in exactly one line, every line containing at least two points. If a linear space S is finite (i.e. has a finite number v of points) and if the size of every line of S is an element of a set $K = \{k_1,\ldots,k_n\}$ of positive integers, we will say that the linear space is an $S(2,K,v)$. Such linear spaces are also known as pairwise balanced designs. When $K = \{k\}$, that is when all lines of S have the same size k, S is a Steiner system $S(2,k,v)$ or, in an equivalent terminology, a 2-$(v,k,1)$ design. The systems $S(2,3,v)$ are usually called Steiner triple systems.

The combinatorial study of finite linear spaces goes back to the 19th century with the pioneering work of Reverend T.P.Kirkman [23] [24] who, among other things, solved in 1847 the problem of existence of Steiner triple systems and gave in 1850 an explicit construction, for every prime p, of an $S(2,p+1,p^2+p+1)$, namely a projective plane of order p. Since that time, much progress has been done, as acknowledged for instance by the bibliography in [15].

The purpose of this paper is to emphasize a few results obtained during the past 10 years. We do not claim for completeness.

2. Existence problems.

If there is a finite linear space $S(2,K,v)$ with $K = \{k_1,\ldots,k_n\}$, the following arithmetical conditions must hold :

$$\text{(i)} \quad v - 1 \equiv 0 \pmod{\alpha(K)}$$

$$\text{(ii)} \quad v(v-1) \equiv 0 \pmod{\beta(K)}$$

where $\alpha(K)$ denotes the g.c.d. of the integers $k_1-1,\ldots,k_n-1$, and $\beta(K)$ the g.c.d. of $k_1(k_1-1),\ldots,k_n(k_n-1)$. Indeed, the lines containing a given point partition the remaining $v-1$ points (which proves (i)), and the $v(v-1)/2$ pairs of points are partitioned by the lines (which proves (ii)).

These necessary conditions of existence are not sufficient : for example, there is no $S(2,\{3,4,5\},8)$ though (i) and (ii) are satisfied. However, as was proved in 1975 by R.M.Wilson [36], they are asymptotically sufficient. More precisely :

<u>Theorem</u> (Wilson) Given K, there is a constant c_K such that a linear space $S(2,K,v)$ exists for all integers $v > c_K$ satisfying (i) and (ii).

For example, an $S(2,\{3,4,5\},v)$ exists for every sufficiently large integer v. An $S(2,\{3,4\},v)$ exists for every sufficiently large integer $v \equiv 0$ or $1 \pmod 3$. An $S(2,3,v)$ exists for every sufficiently large integer $v \equiv 1$ or $3 \pmod 6$.

Let $V(K)$ denote the set of integers v for which there exists an $S(2,K,v)$; an integer $v \in V(K)$ will be called <u>admissible</u>. There are very few sets K for which $V(K)$ is completely determined. Here are three examples :

$$V(\{3,4,5\}) = \mathbb{N} \setminus \{0,2,6,8\}$$

$$V(\{3,4\}) = \{v \mid v \equiv 0 \text{ or } 1 \pmod 3\} \setminus \{0,6\}$$

$$V(\{3\}) = \{v \mid v \equiv 1 \text{ or } 3 \pmod 6\}$$

Note that the mapping $V : K \to V(K)$ is a closure operation on the subsets of $\mathbb{N}_o$. Indeed, it is easily checked that

$$\text{(a)} \quad K \subseteq V(K)$$

$$\text{(b)} \quad K_1 \subseteq K_2 \Rightarrow V(K_1) \subseteq V(K_2)$$

(c) $V(V(K)) = V(K)$

This simple observation is fundamental in the proof of Wilson's theorem. Let us say that K is <u>closed</u> if $V(K) = K$. Then Wilson's theorem asserts that every closed set K contains all sufficiently large integers v for which $v - 1 \equiv 0 \pmod{\alpha(K)}$ and $v(v-1) \equiv 0 \pmod{\beta(K)}$. Another fundamental result is that every closed set K contains a <u>finite</u> subset K' such that $V(K') = V(K)$.

The sets $V(k)$ are known only when $k \leq 5$. More precisely,

An $S(2,3,v)$ exists iff $v \equiv 1$ or $3 \pmod 6$ (Kirkman [23])

An $S(2,4,v)$ exists iff $v \equiv 1$ or $4 \pmod{12}$

An $S(2,5,v)$ exists iff $v \equiv 1$ or $5 \pmod{20}$ (Hanani [18] [19])

Note that, when k is a prime power, (i) and (ii) are equivalent to $v \equiv 1$ or $k \pmod{k(k-1)}$.

By Wilson's theorem, there is an $S(2,6,v)$ for every sufficiently large integer $v \equiv 1, 6, 16$ or $21 \pmod{30}$. Actually, this is known to be true for every such $v > 34366$ and there remain about 400 values of v for which the problem is not solved. If $v < 100$, the non-existence of an $S(2,6,v)$ has been proved for $v = 16, 21$ (easy !) and 36 (equivalent to the non-existence of an affine plane of order 6), and a construction has been found for $v = 1, 6, 31$ (projective plane of order 5), 66 (R.H.F. Denniston in 1978), 76, 91 and 96 (W.H.Mills in 1979, 1974, 1976). Thus the only undecided values of $v < 100$ are now $v = 46, 51, 61$ and 81. Here is Mills' construction of an $S(2,6,91)$: the 91 points are the elements of the additive group $\mathbb{Z}_{91}$ of integers modulo 91 and the 273 lines are all subsets of $\mathbb{Z}_{91}$ of the form

$$\{x,x+1,x+3,x+7,x+25,x+36\}$$

$$\{x,x+5,x+20,x+32,x+46,x+75\}$$

$$\{x,x+8,x+17,x+47,x+57,x+80\}$$

where $x \in \mathbb{Z}_{91}$.

A set of lines partitioning the points of a linear space S is called a parallel class of lines. A linear space whose set of lines can be partitioned into parallel classes is said to be resolvable. In a resolvable $S(2,k,v)$ with $v > 1$, k must divide v. Together with conditions (i) and (ii) above, this implies

$$v \equiv k \pmod{k(k-1)}$$

This necessary condition has been proved to be sufficient for $k = 3$ (Kirkman triple systems) (D.K.Ray-Chaudhuri and R.M.Wilson [30]), for $k = 4$ (H.Hanani, D.K.Ray-Chaudhuri and R.M.Wilson [21]) and for all sufficiently large v (D.K. Ray-Chaudhuri and R.M.Wilson [31]).

Note that the well-known problem proposed in 1850 by Sylvester : "Can fifteen schoolgirls walk out in five rows of three, seven times a week, for a quarter of thirteen weeks, in such a way that any two girls are in the same row just once in each week, and any three just once in each term ?" is equivalent to the problem of partitioning the $\binom{15}{3}$ subsets of size 3 of a set of 15 points into 13 sets of 35 lines, each of them forming a Kirkman triple system. A solution was finally found in 1973 by R.H.F.Denniston [12], with the help of a computer !

3. Isomorphism problems.

Let $N(2,K,v)$ denote the number of pairwise non-isomorphic $S(2,K,v)$.

Much progress has been made during the past decade on the evaluation of $N(2,3,v)$, and so we shall focus on this particular case.

There is, up to isomorphism, a unique $S(2,3,7)$ (namely $PG(2,2)$) and a unique $S(2,3,9)$ (namely $AG(2,3)$). G.Brunel [6] proved in 1897 that

$$N(2,3,13) = 2$$

One of the $S(2,3,13)$ can be constructed as follows : take as points the elements of $\mathbb{Z}_{13}$ and as lines all subsets of the form

$$\{x, x+1, x+4\}$$

$$\{x, x+2, x+7\}$$

where $x \in \mathbb{Z}_{13}$. The other $S(2,3,13)$ is obtained by removing from the above system the lines $\{0,1,4\}$, $\{0,2,7\}$, $\{2,4,9\}$, $\{1,7,9\}$ (which form a Pasch configuration) and replacing them by $\{0,1,7\}$, $\{0,2,4\}$, $\{1,4,9\}$, $\{2,7,9\}$.

The computation of $N(2,3,15)$ was carried out (by hand) in 1917 by F.N.Cole, L.D.Cummings and H.S.White [35] : they found

$$N(2,3,15) = 80$$

The exact values of $N(2,3,v)$ for $v \geq 19$ are unknown. However, there is now some very precise information on the behaviour of $N(2,3,v)$. Indeed, R.M.Wilson [37] proved that, if van der Waerden's permanent conjecture is true,

$$e^{\frac{v^2}{6}(\log v-5)} \leq N(2,3,v) \leq e^{\frac{v^2}{6}(\log v-\frac{1}{2})}$$

for every admissible v (the upper bound is easily established, it is only for the lower bound that van der Waerden's conjecture is needed). The permanent of an $n \times n$ matrix A_n with entries a_{ij} is defined by

$$\text{per } A_n = \sum_{\sigma \in S_n} a_{1\sigma(1)} a_{2\sigma(2)} \cdots a_{n\sigma(n)}$$

where S_n is the symmetric group of degree n.

A_n is called <u>doubly stochastic</u> if its entries are non-negative real numbers and if the sum of all entries in a row or a column of A_n is always 1. The matrix J_n all of whose entries are equal to $\frac{1}{n}$ is doubly stochastic and

$$\text{per } J_n = \frac{n!}{n^n}$$

van der Waerden's permanent conjecture asserts that for every doubly stochastic matrix $A_n \neq J_n$, per $A_n >$ per J_n. This conjecture was proved in 1980 by Egoritsjev (see J.H.van Lint [34]), so that Wilson's bounds for $N(2,3,v)$ are now known to be valid. In particular

$$\log N(2,3,v) \sim \frac{v^2}{6} \log v$$

A linear space is said to be rigid if it has no other automorphism than the identity. C.C.Lindner and A.Rosa [27] proved in 1975 that there is a rigid $S(2,3,v)$ for every admissible $v \geq 15$. Let $N^*(2,3,v)$ denote the number of pairwise non isomorphic rigid $S(2,3,v)$. In 1980, L.Babaï [1] proved that

$$N^*(2,3,v) \sim N(2,3,v)$$

In other words, almost all Steiner triple systems are rigid !

Given any $k > 3$, it is easy to show that

$$N(2,k,v) \leq e^{\frac{(k-2)}{k(k-1)}v^2 \log v}$$

for every admissible v, and Wilson has proved that there is a constant $\alpha_k > 1$ such that

$$\alpha_k^{v^2} \leq N(2,k,v)$$

For $k = 4$, it is known that $N(2,4,v) \geq 2$ for every admissible $v \geq 25$ (A.Brouwer) and that

$$2^{[\frac{v-1}{720}]} \leq N(2,4,v)$$

for every admissible $v \geq 781$ (M.Wojtas [38]).

4. Subspaces.

A subspace of a linear space S is a subset S' of S such that any line passing through two points of S' is entirely contained in S'. Clearly, any intersection of subspaces is a subspace. If X is any subset of S, the subspace generated by X is the intersection of all subspaces containing X.

1) If S is a Steiner system $S(2,k,v)$ and if S' is a subspace of S, S' itself has a structure of Steiner system $S(2,k,v')$. When $S' \neq S$, it is easy to see that

$$v \geq (k-1)v' + 1$$

In the particular case $k = 3$, J.Doyen and R.M.Wilson [16] proved in 1973 that for any two admissible integers v, v' satisfying $v \geqslant 2v' + 1$, there is a Steiner triple system of size v containing a subspace of size v' (another proof has been given by G.Stern and H.Lenz [32]).

The generalization to higher values of k seems to be rather difficult : A.Brouwer and H.Lenz [5] [25] have obtained partial results when $k = 4$ or 5 : for instance, if any two integers $v, v' \equiv 1$ or 4 (mod 12) satisfy $v \geqslant 16v' - 159$, there is an $S(2,4,v)$ containing a subspace of size v'.

2) If S is a Steiner system $S(2,k,v)$ containing at least two lines, S is said to be

(i) a non-degenerate plane if every triangle generates S

(ii) a degenerate plane if there is a triangle which generates S and a triangle which does not

(iii) a Steiner space if there is no triangle generating S.

It is known that there is a non-degenerate plane $S(2,3,v)$ for every admissible $v \geqslant 7$, and a degenerate plane $S(2,3,v)$ for every admissible $v \geqslant 15$ (J.Doyen [13] [14]). A.J.W.Hilton [22] and L.Teirlinck [33] have proved that Steiner spaces $S(2,3,v)$ exist for $v = 15$, 27, 31, 39 and for every admissible $v \geqslant 45$ (except possibly for $v = 51$, 67, 69 and 145) and that they do not exist for $v < 15$ and for $v = 19$, 21, 25, 33, 37 and 43.

For $k \geqslant 4$, the problem is open.

5. Planar spaces.

A planar space is a linear space S provided with a family of distinguished subspaces called planes such that every triangle of S is contained in exactly one plane, every plane containing at least one triangle.

If $k \geqslant 3$, the only known examples of planar $S(2,k,v)$ in which all planes have cardinality $v' < v$ are those in which all planes are projective or all planes are affine. Clearly, a planar space in which all planes are projective is a projective space. F.Buekenhout [7] proved that a planar space in which all planes are affine of order $\geqslant 4$ is an affine space (for affine planes of order 3, this is no longer true, as was shown by M.Hall [17]).

M.Dehon [10] has tried to classify the planar $S(2,3,v)$ in which all planes have the same cardinality $v' < v$ and which satisfy the following regularity condition : for every plane π and every line L disjoint from π, there is a constant number α of planes through L intersecting π in a line. Such a space is necessarily

(i) $PG(d,2)$ with $d \geqslant 3$

(ii) $AG(3,3)$

(iii) an $S(2,3,2(6n+7)(3n^2+3n+1)+1)$ with planes of $6n+7$ points and $\alpha = 1$

(iv) an $S(2,3,171)$ with planes of 15 points and $\alpha = 2$

(v) an $S(2,3,183)$ with planes of 21 points and $\alpha = 7$

(vi) an $S(2,3,1055)$ with planes of 39 points and $\alpha = 4$

Unfortunately, no example of type (iii), (iv), (v) or (vi) is known at present.

In 1971, F.Buekenhout and R.Deherder [8] defined a <u>π-space</u> as a finite planar space S in which all planes are isomorphic to a given linear space π. If S has at least two planes and at least two lines of different sizes, the only known examples are the spaces consisting of $2k$ points lying on two disjoint lines of k points, all the other lines having two points (the planes are degenerate projective planes of $k+1$ points). It is tempting to conjecture that there are no other examples. A.Brouwer [3], A.Delandtsheer [11] and D.Leonard [26] proved that in such a space, each point is in the same number of lines of a given size and the total number of points of the space is uniquely determined by certain arithmetical parameters of π. The smallest unsolved case is a planar space of 47 points in which all planes have 7

points and are isomorphic to

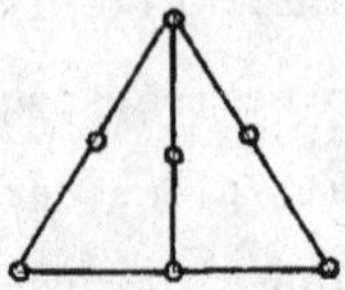

(the lines of size 2 have not been represented).

6. Some classification problems.

1) F.Buekenhout, R.Metz and J.Totten [9] began a classification of finite linear spaces S according to the set T(S) whose elements are the cardinalities of the sets of diagonal points in all quadrangles of S. The set T(S) is a subset of {0,1,2,3}, and so there are 16 possible types. The linear spaces S for which T(S) = {0},{1},{2},{3}, {0,2},{0,3},{1,3},{2,3} have been classified.

An interesting class of linear spaces S consists of those for which $T(S) \subseteq \{0,1\}$: in such a space, there is no Pasch configuration, that is no configuration of 4 lines intersecting in 6 points. These anti-Pasch spaces seem to be rather difficult to classify. Among them, we find the planar spaces in which all planes are isomorphic to AG(2,3), as well as the linear spaces in which all lines of size at least 3 are concurrent.

M.O'Nan [28] has observed that all classical unitals are anti-Pasch (a unital is an $S(2,q+1,q^3+1)$; a classical unital is one obtained from the set of absolute points and non-absolute lines of a unitary polarity in $PG(2,q^2)$). F.Piper [29] conjectured that every anti-Pasch unital is classical. This is clearly true for q = 2 since S(2,3,9) is unique up to isomorphism. For q = 3, A.Brouwer [4] has proved that the only anti-Pasch S(2,4,28) is the classical unital (at the same time, he has constructed at least 87 non isomorphic S(2,4,28)).

A whole family of anti-Pasch Steiner triple systems can be constructed as follows. Let G be a finite multiplicative abelian group of odd order 2n+1. Take as points the 6n+3 elements of $S = G \times \{0,1,2\}$ and as lines the following subsets of S :

(i) $\{(x,0),(x,1),(x,2)\}$ for every $x \in G$

(ii) $\{(x,0),(y,0),(z,1)\}$

$\{(x,1),(y,1),(z,2)\}$

$\{(x,2),(y,2),(z,0)\}$

for every $x, y, z \in G$ such that $x \neq y$ and $xy = z^2$. The $S(2,3,6n+3)$ constructed in this way will be denoted by S_G (it can be shown that S_{G_1} and S_{G_2} are isomorphic if and only if G_1 and G_2 are isomorphic). If $2n+1$ is not divisible by 7, it is easy to check that S_G is anti-Pasch.

We have already seen that $S(2,3,13)$ contains a Pasch configuration. Among the 80 non isomorphic $S(2,3,15)$, only one is anti-Pasch, namely the system obtained by the above construction from a group of order 5.

2) A.Beutelspacher and A.Delandtsheer [2] have recently classified the finite linear spaces S satisfying the following condition :

(*) there is an integer $\alpha \geq 0$ such that for any two disjoint lines L, L' of S and any point p outside $L \cup L'$, there are exactly α lines through p intersecting both L and L'

Such a space is necessarily

(i) a generalized projective space (if $\dim S \geq 4$, all lines of S have size 2)

(ii) an affine plane, an affine plane with one point at infinity or a punctured projective plane

(iii) the linear space obtained from $PG(2,2)$ by breaking one of its lines into 3 lines of size 2

A.Delandtsheer (unpublished) has also classified completely the finite linear spaces S satisfying :

(**) there is an integer $\beta > 0$ such that for any two disjoint lines L, L' of S and any point p outside $L \cup L'$, there are exactly β lines through p disjoint from $L \cup L'$.

BIBLIOGRAPHY

1. L.BABAI : Almost all Steiner triple systems are asymmetric, Annals of Discrete Math. 7(1980), 37-39.
2. A.BEUTELSPACHER and A.DELANDTSHEER : A common characterization of finite projective spaces and affine planes (to appear).
3. A.BROUWER : On the nonexistence of certain planar spaces (to appear).
4. A.BROUWER : Some unitals on 28 points and their embeddings in projective planes of order 9 (this volume).
5. A.BROUWER und H.LENZ : Unterräume von Blockplänen. In: Contributions to Geometry, Birkhäuser 1979, pp. 383-389.
6. G.BRUNEL : Sur les deux systèmes de triades de treize éléments, J.Math.Pures Appl. (5) 7(1901), 305-330.
7. F.BUEKENHOUT : Une caractérisation des espaces affins basée sur la notion de droite, Math.Z.111(1969), 367-371.
8. F.BUEKENHOUT et R.DEHERDER : Espaces linéaires finis à plans isomorphes, Bull.Soc.Math.Belg. 23 (1971), 348-359.
9. F.BUEKENHOUT, R.METZ and J.TOTTEN : A classification of linear spaces based on quadrangles I, Simon Stevin 52(1978), 31-45.
10. M.DEHON : Planar Steiner triple systems, J.Geometry 12 (1979), 1-9.
11. A.DELANDTSHEER : Finite planar spaces with isomorphic planes (to appear).
12. R.H.F.DENNISTON : Sylvester's problem of the 15 schoolgirls, Discrete Math. 9 (1974), 229-233.
13. J.DOYEN : Sur la structure de certains systèmes triples de Steiner, Math.Z. 111(1969), 289-300.
14. J.DOYEN : Systèmes triples non engendrés par tous leurs triangles, Math.Z. 118(1970), 197-206.
15. J.DOYEN and A.ROSA : An updated bibliography and survey of Steiner systems, Annals of Discrete Math. 7(1980), 317-349.
16. J.DOYEN and R.M.WILSON : Embeddings of Steiner triple systems, Discrete Math. 5(1973), 229-239.
17. M.HALL, Jr : Automorphisms of Steiner triple systems, IBM J.Res. Develop. 4(1960), 460-472.
18. H.HANANI : The existence and construction of balanced incomplete block designs, Ann.Math.Statist. 32(1961), 361-386.

19. H.HANANI : A balanced incomplete block design, Ann.Math.Statist. 36(1965), 711.

20. H.HANANI : Balanced incomplete block designs and related designs, Discrete Math. 11(1975), 255-369.

21. H.HANANI, D.K.RAY-CHAUDHURI and R.M.WILSON : On resolvable designs, Discrete Math. 3(1972), 343-357.

22. A.J.W.HILTON : On the Szamkolowicz-Doyen classification of Steiner triple systems, Proc.London Math.Soc (3) 34 (1977), 102-116.

23. T.P.KIRKMAN : On a problem in combinations, Cambridge and Dublin Math. J.2(1847), 191-204.

24. T.P.KIRKMAN : On the perfect r-partitions of $N = r^2 - r + 1$, Transactions of the Historic Society of Lancashire and Cheshire 9 (1856-57), 127-142.

25. H.LENZ : Embedding block designs into larger ones, Preprint 45, Freie Universität Berlin, 1977.

26. D.LEONARD : Finite π-spaces (to appear).

27. C.C.LINDNER and A.ROSA : On the existence of automorphism free Steiner triple systems, J.Algebra 34 (1975), 430-443.

28. M.O'NAN : Automorphisms of unitary block designs, J.Algebra 20 (1972), 495-511.

29. F.PIPER : Unitary block designs, in : Graph Theory and Combinatorics (R.J.Wilson, ed.), Research Notes in Mathematics 34.

30. D.K.RAY-CHAUDHURI and R.M.WILSON : Solution of Kirkman's schoolgirl problem, Proc.Symp.Pure Math. 19 (Amer.Math.Soc., Providence RI, 1971), 187-203.

31. D.K.RAY-CHAUDHURI and R.M.WILSON : The existence of resolvable block designs, in : A survey of combinatorial theory (North Holland, Amsterdam, 1973), 361-375.

32. G.STERN and H.LENZ : Steiner triple systems with given subspaces, another proof of the Doyen-Wilson theorem, Boll.Un.Mat.Ital. A (5) 17(1980), 109-114.

33. L.TEIRLINCK : On Steiner spaces, J.Combinatorial Theory (A) 26 (1979), 103-114.

34. J.H.van Lint : Notes on Egoritsjev's proof of the van der Waerden conjecture, Memorandum 1981-01, Eindhoven University of Technology 1981.

35. H.S.WHITE, F.N.COLE and L.D.CUMMINGS : Complete classification of the triad systems on fifteen elements, Memoirs Nat.Acad.Sci. USA 14, 2nd memoir (1919), 1-89.

36. R.M.WILSON : An existence theory for pairwise balanced designs, III Proof of the existence conjectures, J.Combinatorial Theory (A) 18 (1975), 71-79.

37. R.M.WILSON : Non isomorphic Steiner triple systems, Math.Z. 135 (1974), 303-313.

38. M.WOJTAS : On non isomorphic balanced incomplete block designs B(4,1,v), Colloq.Math. 35(1976), 327-330.

ON DESIGNS

D.R. HUGHES

In this little paper we shall give some of the basic facts about designs, and pass on to a few selected topics that are more advanced. Design Theory today has become such a vast subject that a "review" of all its elements and developments would be extremely long. So we have concentrated on displaying some of the flavour, omitting many important topics. This has sometimes caused anguish for the author: we hope that it does not meet with excessive disapproval from the experts! A complete bibliography would be almost as long as the paper, so we have chosen the opposite course and given none. For many of the older theorems, this lack is filled by Dembowski's "Finite Geometries", and when its revised edition appears soon the situation will be even more satisfactory. But there are results in the paper which have not appeared anywhere, although we hope that they will in the near future.

We use a fairly consistent terminology, not all of which is universally standard, though none of it should be shocking. It is the terminology that will be used in the soon-to-appear (but long-promised) book on design theory by Piper and the present author. (Hence the word "group" in "group-divisible" is not only redundant but also somewhat illogical to an algebraist, so we simply suppress it, etc.) We have tried to be precise and yet informal, and trust that the paper can be read in that spirit.

1. INTRODUCTION

Let $\underline{P}$ and $\underline{B}$ be finite sets whose elements are called points and blocks, respectively, and let $\underline{I}$ be a subset of $\underline{P} \times \underline{B}$. We write PIy if $(P,y) \in \underline{I}$, and say that P is incident with y or P is on y, or the dual of either of these (or indeed, use any other unambiguous "geometric" terminology). Then $\underline{S} = (\underline{P},\underline{B},\underline{I})$ is a structure. We can identify the set of points on a block with the block itself, as long as we allow that $\underline{S}$ might contain "repeated" blocks or even empty blocks, but we shall be interested in structures without repeated (or empty) blocks.

Definition : The dual $\underline{S}^T$ of $\underline{S}$ is the structure $(\underline{B},\underline{P},\underline{I}')$ where $(y,P) \in \underline{I}'$ if and only if $(P,y) \in \underline{I}$.

Definition : A structure $\underline{S}$ is connected if given any two of its elements, say a,b, there is a set $a_1, a_2, \ldots, a_n$, $n \geq 0$, of elements of $\underline{S}$ such that

$$a I a_1 I a_2 \ldots I a_n I b.$$

Now let r_p be the number of blocks on the point P and k_y the number of points on the block y.

Definition : If $r_p = r$ for all $P \in \underline{P}$, then $\underline{S}$ is regular (with point size r); if $k_y = k$ for all $y \in \underline{B}$, then $\underline{S}$ is uniform (with block size k).

Definition : A structure $\underline{S}$ is proper if there is a block $y \in \underline{S}$ such that $1 < k_y < |\underline{P}|$.

We usually write $v = |\underline{P}|$, $b = |\underline{B}|$.

Definition : A flag of $\underline{S}$ is a pair $(P,y) \in \underline{I}$.

Definition : If $\underline{S}$ is uniform and proper and has no repeated blocks, then $\underline{S}$ is a design. A design with block size k such that every k-set of points is a block is trivial.

Lemma 1 : If $\underline{S}$ is a uniform and regular structure then $vr = bk$.

Proof : Count flags. ¤

Definition : Let $t > 0$ be an integer. If there is an integer $\lambda > 0$ such that any t-set of points of $\underline{S}$ is in exactly λ common blocks, then $\underline{S}$ is a t-structure.

Definition : If $\underline{S}$ is a uniform t-structure with block size k, and λ blocks on t points, then $\underline{S}$ is a t-structure for (v,k,λ); if in addition it is a design, then it is a t-design for (v,k,λ), or merely a t-(v,k,λ).

Lemma 2 : If $\underline{S}$ is a t-structure for (v,k,λ) and if i is an integer, $0 \leq i < t$, then any i-set of points of λ is on exactly λ_i blocks, where

$$\lambda_i = \lambda \frac{(v-i)(v-i-1)\ \dots\ (v-t+1)}{(k-i)(k-i-1)\ \dots\ (k-t+1)} .$$

Proof : Fix an i-set $\underline{C}$ of points. Then count in the two different ways the numbers of pairs $(\underline{A},y)$, where $\underline{A}$ is a t-set of points, $\underline{C} \subseteq \underline{A}$, and y is a block on $\underline{A}$. ¤

Lemma 2 is, in fact, a sort of non-existence theorem, since each λ_i must be integral; it also yields an expression for $b = \lambda_0$, and implies that $r = \lambda_1$ is a constant, so $\underline{S}$ is regular.

If $\underline{S}$ is a structure with v points and b blocks, let $A = (a_{py})$ be a $v \times b$-matrix, whose rows are indexed by the points and whose columns are indexed by the blocks, and $a_{py} = 0$ or 1 according as P is not, or is, on y. A is an <u>incidence matrix</u> for $\underline{S}$. We write $\underline{j}$ for a (row) vector of all + 1's, and $J_{n,m}$ for an $n \times m$ matrix all of whose entries are +1 (we write J_m for $J_{m,m}$, and even J for $J_{n,m}$, if n and m are clear). Clearly :-

<u>Lemma 3</u> : $\underline{j}\, A = (k_{y_1}, k_{y_2}, \ldots, k_{y_b})$ and

$$A\underline{j}^T = (r_{P_1}, r_{P_2}, \ldots, r_{P_v}).$$

Incidence matrices have many useful applications. For instance, if $\underline{S}$ is a 2-structure for (v,k,λ), then

(a) $AA^T = nI + \lambda J$, where $n = r-\lambda (=\lambda_1 - \lambda)$,

(b) $AJ = rJ$ and $JA = kJ$,

and it is easy to see that a (0,1)-matrix A satisfying (a) and (b) is an incidence matrix of a 2-structure for (v,k,λ). It is also easy to compute the determinant of $nI + \lambda J$ (e.g., subtract the top row from every other, then add every column, but the first, to the first), and see that

(c) $\det(nI+\lambda J) = n^{v-1}(n+v\lambda)$.

From this one proves

<u>Theorem 4</u> : (Fisher's Inequality). If $\underline{S}$ is a proper uniform 2-structure with v points and b blocks, then $b \geq v$.

Proof : The rank of $nI + \lambda J$ is v, so the rank of A is v as well. ¤

Definition : A structure is square if $v = b$, and a square 2-design is a symmetric design. A square structure is non-singular if one (and hence all) of its incidence matrices is non-singular (over the real field).

Theorem 5 : A proper square 2-structure is uniform if and only if it is regular, and in either case it is then a symmetric design.

Theorem 6 : Let S be a 2-design for (v,k,λ). Then the following are equivalent :-

(a) $b = v$;

(b) $k = r$;

(c) S^T is a 2-structure;

(d) S^T is a 2-design for (v,k,λ).

The proofs of Theorems 5 and 6 are well-known but long and we omit them. Similarly we omit the (much less elementary) proof of the famous Bruck-Ryser-Chowla Theorem.

Theorem 7 : Let S be a symmetric design for (v,k,λ). Then :-

(a) if v is even, then $n = k-\lambda$ is a square;

(b) if v is odd, then there exist integers x,y,z, not all zero, such that

$$x^2 = (k-\lambda)y^2 + (-1)^{\frac{v-1}{2}}\lambda z^2.$$

It is in fact true that while infinitely many symmetric designs with $\lambda = 1$ are known, for each fixed $\lambda > 1$ only finitely many symmetric designs are known at present. This may be part of a larger problem : no single set of possible parameters (v,k,λ) for a symmetric design has ever been shown to be impossible, except those rejected by Theorem 7. Thus the known symmetric design for $\lambda = 1$ all have $n(=k-1)$ a power of a prime (and every such prime-power is possible), but Theorem 7 rejects only half of the remaining values of n and for no single one of these (the smallest being $n = 10$) do we know whether a design exists, or not.

Among the most interesting and important aspects of Design Theory is the connection to Group Theory. An <u>automorphism</u> of a structure has the obvious definition, and by Aut $\underline{S}$ we shall mean the full automorphism group of $\underline{S}$. If $\alpha \in$ Aut $\underline{S}$ and A is an incidence matrix for $\underline{S}$, then α induces permutations on the points and blocks of $\underline{S}$; representing these by permutation matrices P,Q respectively, then $PAQ^{-1} = A$.

<u>Lemma 8</u> : If $\underline{S}$ is square and non-singular, and $\alpha \in$ Aut $\underline{S}$, then α has the same cycle structure on points and on blocks.

<u>Proof</u> : From $PAQ^{-1} = A$, we have $A^{-1}PA = Q$, so P and Q have the same characteristic equations. The characteristic equation of a permutation matrix with a_i i-cycles is

$$\Pi(x^i-1)^{a_i},$$

and equating the two expressions of this sort for P and Q, by repeatedly comparing the "largest" cyclotomic polynomials on the two sides we see that the number of i-cycles for P and Q are the same, for all i. ⌑

Theorem 9 : If $\underline{S}$ is square and non-singular and G is a subgroup of Aut $\underline{S}$ then G has equally many point and block orbits.

Proof : If H is a permutation group and for $h \in H$, $\chi(h)$ is the number of fixed symbols of h, then it is well-known that

$$t|H| = \sum_{h \in H} \chi(h)$$

where t is the number of orbits of H. Applying this to G, in its two representations in points and blocks, and using Lemma 8, our result follows. ⌑

Note that Theorem 9 does not imply that G has the same number of fixed points and fixed blocks. E.g., the translation group with a fixed axis of a classical projective plane of order n has one fixed block (= line) and n+1 fixed points.

Definition : An automorphism group $G \leq$ Aut $\underline{S}$ of a square structure $\underline{S}$ is a Singer group if it is regular on points and on blocks.

In view of Theorem 9, if $\underline{S}$ is non-singular it would suffice in the definition of a Singer group to demand that G be regular on points (or on blocks).

Theorem 10 : If G is a Singer group for $\underline{S}$, if P is a (fixed) point, y a (fixed) block of $\underline{S}$, and if $D = D(P,y) = \{g \in G | P^g \text{ on } y\}$, then we can represent $\underline{S}$ as :-

$$Q = P^g \leftrightarrow (g)$$
$$z = y^h \leftrightarrow [h]$$

where (g) is on [h] if and only if $g \in Dh$. (So the points of S "are" the elements of G and the blocks of S "are" the subsets Dh.)

The proof is obvious.

The representation of Theorem 10 is even more useful if S is a symmetric design. We call the set D of Theorem 10 a difference set for G in this case, and have :-

Theorem 11 : If G is a Singer group for the symmetric design S, whose parameters are (v,k,λ), and if D is a difference set for G, then

(a) $g \in G,\ g \neq 1 \Rightarrow g = d_1 d_2^{-1}$ for precisely λ pairs $d_1, d_2 \in D$;

(b) $g \in G,\ g \neq 1 \Rightarrow g = d_1^{-1} d_2$ for precisely λ pairs $d_1, d_2 \in D$.

Conversely if G is a group of order v and D is a subset of k elements in G for which either (a) or (b) holds, then G is a Singer group of a symmetric design with parameters (v,k,λ) and D is one of its difference sets (in this representation).

We omit the proof, which is not difficult : for the first part, consider the λ blocks y^h on the two points P and P^g, etc. For the converse, S is constructed as in Theorem 10.

Singer groups are known to exist for the symmetric designs whose points and blocks are the points and hyperplanes of the classical projective geometry PG(n,q), $n \geq 2$, and also for infinitely many other examples. But L.J. Dickey and the author have conducted a computer search for abelian Singer groups for symmetric designs with $\lambda = 2$ and $k \leq 5000$ and shown that none can exist in this range if $k > 9$. It is also an old conjecture that if $\lambda = 1$, i.e., $\underline{S}$ is a projective plane, then Singer groups exist if and only if the design is a classical projective plane, and considerable theoretical and computational evidence for this conjecture exists. The chief tools in all these studies are the multiplier theorems.

Definition : Let $G \leq \text{Aut } \underline{S}$ be an automorphism group of the structure $\underline{S}$, and let N be the normalizer of G in Aut $\underline{S}$. Then $M = N/G$ is the multiplier group of G, and its elements are multipliers of G.

In the case of Singer groups, the following is basic :-

Theorem 12 : If G is a Singer group of $\underline{S}$, and N is the normalizer of G in Aut $\underline{S}$, then the multiplier group M is isomorphic to any stabilizer N_p, where P is a point of $\underline{S}$. An automorphism ϕ of the (abstract) group G induces an element of N_p if and only if $D^\phi = Da$, for some $a \in G$, where $D = D(p,y)$ is a difference set (and y is an arbitrary block of $\underline{S}$).

Proof : The first sentence is clear, since N is a transitive group with a regular normal subgroup G. The second sentence is proved as follows :-

If $b \in N_p$, then (where $D = D(P,y)$) y^b is a block, so $y^b = y^a$, where $a \in G$. Then for $d \in D$, $b^{-1}d\, b \in G$, and $P^{b^{-1}db} = P^{db}$ is on $y^b = y^a$; hence $b^{-1}d\, b \in Da$, so $b^{-1} D\, b$ $(=D^b) = Da$. Conversely, if $\phi \in \text{Aut } G$ and $D^\phi = Da$, define $\overline{\phi}$ on $\underline{S}$ as follows :-

$$(p^g)^{\overline{\phi}} = p^{g^{\phi}}$$

$$(y^g)^{\overline{\phi}} = y^{a.g^{\phi}} \quad .$$

Then it is easy to see that $\overline{\phi} \in \text{Aut } \underline{S}$, and in fact $\overline{\phi} \in N$ and finally $\overline{\phi} \in N_p$. ¤

There are many multiplier theorems for abelian Singer groups, and we give only the (historically) first of them, due to Marshall Hall (but the others are, broadly, similar in flavour).

Theorem 13 : Let G be an abelian Singer group for a symmetric design $\underline{S}$ with parameters (v,k,λ). Let p be a prime satisfying

(a) $(p,v) = 1$,

(b) $p|n = k - \lambda$,

(c) $p > \lambda$.

Then the automorphism $g \to g^p$ of G is a multiplier of G.

It is widely conjectured that condition (c) of Theorem 13 is not necessary, and the other multiplier theorems are essentially attempts to remove it. Theorem 13 is, however, very powerful and permits both construction and non-existence applications. First we note that if D is a difference set for the abelian group G, then Dg is a difference set as well (and even for the same design). If $(v,k) = 1$ and $d_0 = \Pi d$, (over all $d \in D$), then there is an element $g \in G$ such that $g^{-k} = d_0$; replace D by Dg, and then the product of all the elements in Dg is $1 \in G$, so we assume $d_0 = 1$ above. Then if ϕ is any multiplier of G, so $D^{\phi} = Da$, it is easy to see that $a = 1$. This is an instance of a more general theorem about choosing D so that all multipliers fix it, but using this little result (i.e., $(v,k)=1$), we have :-

Example 1 : If a symmetric (11,5,2) exists with a Singer group, we write the group additively as C_{11}. Then $p = 3$ satisfies Theorem 13, so if $x \in C_{11}$ and $x \in D$, then $3x, 4x, 5x, 9x \in D$. In fact $D = \{1,3,4,5,9\}$ is a difference set for C_{11}.

Example 2 : If a projective plane of order 10 exists with an abelian Singer group, again we represent the group as C_{111}. Both $p = 2$ and $p = 5$ satisfy the condition of Theorem 13, so if $d \in D$, then $2d, 4d, 5d \in D$. But $2d - d = 5d - 4d$, and thus Theorem 11, either $d = 0$ or $4d = d$. So D cannot contain more than the three elements of C_{111} satisfying $3d = 0$, and hence D cannot exist (for it must contain eleven elements).

2. HIGHER VALUES OF t

Although much attention has been given to t-designs with $t \geq 2$, and there would seem to be the situation with the most "structure", no non-trivial 6-design has ever been found. Certain 4- and 5-designs are of considerable importance and we say something about them here.

Definition : Let $\underline{S}$ be a structure and P a point of $\underline{S}$. Then $\underline{S}_P$, the (internal) restriction or the residue of $\underline{S}$ at P, is the set of all blocks incident with P and of all points, except P, on those blocks; $\underline{S}_y$, for y a block, is defined similarly (i.e. dually).

Theorem 14 : If $\underline{S}$ is a t-(v,k,λ), then $\underline{S}_P$ is a $(t-1)$-$(v-1,k-1,\lambda)$.

The proof is immediate.

Conversely, we define $\underline{S}^*$ to be an extension of the t-design $\underline{S}$ if $\underline{S}^*$ is a (t+1)-design and $\underline{S}^*_P \cong \underline{S}$ for some point P in $\underline{S}^*$.

Lemma 15 : Let $\underline{S}$ be a 2-design for (v,k,λ) and X,Y,Z three distinct points of $\underline{S}$. If m_0 is the number of blocks of $\underline{S}$ that contain none of X,Y,Z and m_3 is the number of blocks of $\underline{S}$ that contain all three of X,Y,Z, then $m_0 + m_3 = b - 3r + 3\lambda$ (where b and r are the number of blocks of $\underline{S}$ and the number of blocks on a point, respectively).

Proof : If m_{XY} is the number of blocks on X and Y, but not on Z, then $m_{XY} + m_3 = \lambda$, the total number of blocks on X and Y. So $m_{XY} = \lambda - m_3$. Hence $m_{XY} = m_{YZ} = m_{XZ} = \lambda - m_3$. Similarly, if m_X is the number of blocks on X but not on Y or Z, then $m_X + m_{XY} + m_{XZ} + m_3 = r$ so $m_X = m_Y = m_Z = r - 2\lambda + m_3$. Then $b = m_0 + m_X + m_Y + m_Z + m_{XY} + m_{YZ} + m_{XZ} + m_3$ so $m_0 + m_3 = b - 3r + 3\lambda$. ¤

Theorem 16 : Let $\underline{S}$ be a 2-design for $(2k+1,k,\lambda)$. Then $\underline{S}$ has an extension $\underline{S}^*$, which is a 3-design for $(2k+2,k+1,\lambda)$.

Proof : If $v = 2k+1$, then from Lemma 2, $b = 2\lambda(2k+1)/(k-1)$, $r = 2\lambda k/(k-1)$, and so $b - 3r + 3\lambda = \lambda$. We construct $\underline{S}^*$ as follows : its points are the points of $\underline{S}$, plus one new point ∞; its blocks are (as point sets)

(a) $y \cup \infty$, where y is a block of $\underline{S}$,

(b) the complements y^c in $\underline{S}$ of blocks y of $\underline{S}$.

Then $\underline{S}^*$ is uniform of block size k + 1 and, from Lemma 15, it is seen that $\underline{S}^*$ is a 3-design. Clearly $\underline{S}^*_\infty = \underline{S}$. ◻

The extension $\underline{S}^*$ of Theorem 16 may not be unique; e.g. there is a 2-(9,4,3) with two non-isomorphic extensions to 3-designs for (10,5,3). But if $\underline{S}$ is <u>symmetric</u>, then it can easily be shown that $\underline{S}^*$ is the only extension (up to isomorphism) of $\underline{S}$. This class of symmetric designs have been much studied and are closely related to many other combinatorial problems.

<u>Definition</u> : If $\underline{S}$ is a symmetric design with $v = 2k + 1$, then $\underline{S}$ is a <u>hadamard 2-design</u>. The extension $\underline{S}^*$ of Theorem 16 of a hadamard 2-design is a <u>hadamard 3-design</u>.

It is, in fact, immediate that $k = 2\lambda + 1$ for a hadamard 2-design, so it has parameters $(4\lambda+3, 2\lambda+1, \lambda)$. Hadamard 2-designs exist for infinitely many values of λ, but it is an open question as to whether they exist for all λ. (In fact the set of values of λ for which a hadamard 2-design is known to exist has density 0 in the positive integers, as far as the author knows, even though there are very few values less than, say, 1000 for which none is known.)

Theorem 16 actually is a special case of another result :-

<u>Theorem 17</u> : If $\underline{S}$ is a 2s-design for $(2k+1, k, \lambda)$ then $\underline{S}$ has an extension.

From this, we can construct an interesting design as follows. Let $\underline{L}$ be the symmetric (11,5,2) of Example 1 in Section 1. Then $\underline{L}^*$, the extension of $\underline{L}$, is a 3-(12,6,2). (In fact $\underline{L}$ is unique, which is easy to prove simply by utilizing the smallness of the parameters; so every residue $\underline{L}^*_p$ of $\underline{L}^*$ is isomorphic to $\underline{L}$, and from this it will follow that Aut $\underline{L}^*$ is transitive on points. Then Aut $\underline{L}^*$ can be shown to be a simple group of order $12 \cdot 11 \cdot 10 \cdot 6$, 3-transitive of degree 12, and abstractly isomorphic to the Mathieu group M_{11}.) Now $\underline{L}^*$ permits the construction of another design.

Lemma 18 : Let $\underline{H}$ be a hadamard 3-design for $(4\lambda+4, 2\lambda+2, \lambda)$. Define the structure $\underline{S}$ as follows : points of $\underline{S}$ are pairs (y,y'), where y is a block of $\underline{H}$ and y' is its complement; blocks of $\underline{S}$ are pairs $[P,Q]$, where P,Q are distinct points of $\underline{H}$; (y,y') is on $[P,Q]$ in $\underline{S}$ if P,Q are both on y, or both on y', in $\underline{H}$.

Then $\underline{S}$ is a $2-(4\lambda+3, 2\lambda+1, 2\lambda(\lambda+1))$.

Proof : The values of v and k for $\underline{S}$ come from straight-forward counting. That $\underline{S}$ has no repeated blocks is easy. If (y,y') and (z,z') are distinct points of $\underline{S}$, then (as point sets) $y \cap z$ is not empty. If X is a point in $y \cap z$, then $\underline{H}_X$ is a hadamard 2-design in which y,z are blocks, so y and z meet λ times in $\underline{H}_X$. Then $|y \cap z| = \lambda + 1$, and similarly, $|y \cap z'| = \lambda + 1$, etc. A block $[P,Q]$ is on (y,y') and (z,z') if and only if P,Q lie together in one of the four sets $y \cap z$, $y \cap z'$, $y' \cap z$, $y' \cap z'$, and since each has $\lambda + 1$ points, there are $4((\lambda+1)\lambda/2) = z(\lambda+1)$ blocks on the two points of $\underline{S}$. □

In the special case $\lambda = 2$, the design $\underline{S}$ above is not only a $2-(11,5,12)$ but is in fact a 4-design for $(11,5,1)$. Hence, by Theorem 17, $\underline{S}$ has an extension to $\underline{S}^*$, a $5-(12,6,1)$. It turns out that Aut $\underline{S}^*$ is M_{12}, and in fact an easy and elementary combinatorial proof of the existence of M_{11} and M_{12}, and their simplicity, can be put together this way. (It is also true that the design $\underline{S}$ of Lemma 18 can never be a 3-design if $\lambda \neq 2$; this can be proved directly using Lemma 2.)

The other t-designs for t = 4 and 5 that concern us require more work, but the construction is still fairly elementary. We sketch the main idea. There is a unique projective plane of order 4; i.e., a unique symmetric design $\underline{P}$ with parameters (21,5,1). The automorphism group of $\underline{P}$ is PΓL(3,4), and contains normal subgroups PGL(3,4) and PSL(3,4) such that $|P\Gamma L : PGL| = 2$, $|PGL : PSL| = 3$ (we abbreviate PΓL(3,4), etc. in the obvious way). An <u>oval</u> in $\underline{P}$ is a set of six points, no three collinear (not on a "block"); elementary counting shows that $\underline{P}$ contains 168 ovals, that PGL is transitive on ovals and that PSL has three orbits of 56 ovals each. A <u>Baer subplane</u> of $\underline{P}$ is a subset of 7 points and 7 lines which is a 2-(7,3,1); we abbreviate Baer subplane by Bsp. It is not difficult to show that $\underline{P}$ contains 360 Bsp's, that PGL is transitive on these and that PSL has three orbits of 120 Bsp's each. A <u>quadrangle</u> of $\underline{P}$ is a set of 4 points, no 3 collinear; PGL is, of course, transitive on these, while PSL has 3 orbits of quadrangles. Let the PSL orbits of ovals be $\overline{\theta}_1, \overline{\theta}_2, \overline{\theta}_3$ of Bsp's be $\overline{B}_1, \overline{B}_2, \overline{B}_3$, and of quadrangles be $\overline{Q}_1, \overline{Q}_2, \overline{Q}_3$. We can choose the subscripts so that :

the oval $\theta \in \overline{\theta}_i \Leftrightarrow$ all quadrangles in θ are in $\overline{Q}_i$

the Bsp $\underline{B} \in \overline{B}_i \Leftrightarrow$ all quadrangles in $\underline{B}$ are in $\overline{Q}_i$.

Both of these follow from the crucial observation that the stabilizer in PSL of an oval (or of a Bsp) is transitive on the quadrangles in that oval (or in that Bsp).

Then we define a structure $\underline{M}_{24}$ as follows :

Points : points of $\underline{P}$ and 3 new points X_1, X_2, X_3.

Blocks (as point-sets) :

(a) $y \cup X_1 \cup X_2 \cup X_3$, where y is a line of $\underline{P}$;

(b) $\theta \cup X_i \cup X_j$, where θ is an oval in $\underline{P}$, $\theta \in \overline{\theta}_s$, and $\{s,i,j\} = \{1,2,3\}$;

(c) $\underline{B} \cup X_i$, where $\underline{B}$ is a Bsp in $\underline{P}$, $\underline{B} \in \overline{B}_i$;

(d) $y * z$, which is the symmetric difference of the distinct lines y,z of $\underline{P}$, (i.e., the 8 points which are on y or z, but not both).

Then all of the following can be proved by elementary (but detailed and careful) combinatorial arguments :

<u>Theorem 19</u> : $\underline{M}_{24}$ is a 5-(24,8,1) and up to isomorphism is the unique design with those parameters. Its successive restrictions, $\underline{M}_{23}$ and $\underline{M}_{22}$, are also unique with their parameters (i.e., 4-(23,7,1) and 3-(22,6,1)). M_{24} = Aut $\underline{M}_{24}$ and M_{23} = Aut $\underline{M}_{23}$ are simple groups, respectively 5 and 4-fold transitive on points; $\overline{M}_{22}$ = Aut $\underline{M}_{22}$ has a normal simple subgroup M_{22} of index 2, and M_{22} is 3-transitive on points. Finally $|M_{22}| = 22\cdot 21\cdot 20\cdot 48$, $|M_{23}| = 23\cdot|M_{22}|$, $|M_{24}| = 24\cdot|M_{23}|$.

The groups $M_{24}, M_{23}, M_{12}, M_{11}$ are the only 4-transitive groups known other than alternating and symmetric groups and (in the light of the classification of finite simple groups) are presumably the only such finite groups which can exist.

The designs constructed are very useful. For instance a very straight-forward construction of the Golay codes can be given by considering the column space over GF(2) of incidence matrices for M_{23}

and M_{24}. By various tricks M_{22} can be used to construct other designs - for instance, there are 16 blocks in M_{22} that do not meet a fixed block y and together with the 16 points not on y, this gives a symmetric design for (16,6,2); the 56 blocks not through a fixed point P form a symmetric design (56,11,2) whose points and blocks are both represented by the 56 ovals, with incidence given by equality or having no point in common. (These statements are very easy to demonstrate.)

3. 1-DESIGNS

A 1-design has very little "structure" : a uniform and regular structure without repeated blocks. But many special 1-designs have great importance. For instance, generalized quadrangles are 1-designs, whose blocks are called "lines", satisfying :

(a) 2 points are on 0 or 1 line;

(b) if P is a point and y is a line, P not on y, then there is a unique line on P which is also on a point of y.

The theory of generalized quadrangles is too rich for us to deal with them even superficially here; they arise naturally in many settings, classical projective geometry and the study of finite simple groups in particular (but also in non-classical finite translation planes, for instance). They have been generalised in a number of very important ways.

Another important class of 1-designs are given by :

Definition : A square 1-design $\underline{S}$ is a partial symmetric design (a PSD) if there exist integers $\lambda_1, \lambda_2 \geq 0$, such that

(a)   2 points of $\underline{S}$ are on λ_1 or λ_2 common blocks;

(b)   2 blocks of $\underline{S}$ contain λ_1 or λ_2 common points and all such that

(c) $\underline{S}$ is connected.

(Note that (c) is superfluous if both λ_i are non-zero;   also if $\lambda_1 = \lambda_2$ then $\underline{S}$ is a symmetric design.)

<u>Definition</u> : A PSD is a <u>semi-symmetric design</u> (SSD) if $\lambda_1 = 0$.

<u>Lemma 20</u> : Let $\lambda > 1$ and let $\underline{S}$ be a proper structure satisfying

(a)   2 points of $\underline{S}$ are on 0 or λ blocks;

(b)   2 blocks of $\underline{S}$ meet in 0 or λ points;

(c) $\underline{S}$ is connected.

Then $\underline{S}$ is a SSD.

<u>Proof</u> : Let P be a point and y a block of $\underline{S}$, and let m be the number of flags (X,z), where X is on y, P is on z, $X \neq P$, $x \neq y$.   There are $k_y - 1$ choices of X on y, and since each is joined to P by y, it is joined by $\lambda - 1$ blocks $z \neq y$.   So $m = (k_y - 1)(\lambda - 1)$.   But there are $r_p - 1$ choices of z on P and each meets y in P, hence in $\lambda - 1$ points $X \neq P$, and so $m = (r_p - 1)(\lambda - 1)$.   Thus $r_p = k_y$, and so, by connectivity, $\underline{S}$ is uniform and regular with $k = r$.   Counting all the flags in $\underline{S}$ implies

that $\underline{S}$ is square. If $\underline{S}$ has repeated blocks, then it easily follows that $k = \lambda$ and hence $v = k$, so $\underline{S}$ is not proper. ¤

SSD's are among the most interesting of PSD's and certainly have very "attractive" properties. Among SSD's the class that has been studied the most are those with $\lambda = 2$: these are called <u>semi-biplanes</u>. They are the finite geometries associated with the Buekenhout diagram

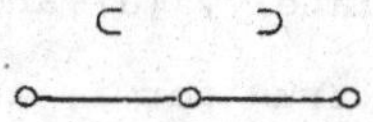

In the light of Lemma 20, we speak of a SSD for $(v,k,(\lambda))$.

<u>Theorem 21</u> : If $\underline{S}$ is a SSD for $(v,k,(\lambda))$, with $\lambda > 1$, then

$$k(k-1)/\lambda + 1 \leq v \leq (\lambda-1)2^{k-1}/\binom{k-2}{\lambda-2}.$$

In addition, $\lambda | k(k-1)$, and $v = k(k-1)/\lambda + 1$ if and only if $\underline{S}$ is a symmetric design.

<u>Proof</u> : If P is a point in $\underline{S}$, we count the number M_2 of points in $\underline{S}_P$ as follows : the number of flags (X,y), where X is on y, $X \in \underline{S}_P$, and y is on P is $M_2 \cdot \lambda$ and is also $k(k-1)$, since there are k blocks y and k-1 choices of X on y. So $M_2 = k(k-1)/\lambda$, and $\lambda | k(k-1)$. Clearly $\underline{S}$ is a 2-design (hence symmetric) if and only if $v = k(k-1)/\lambda+1$.

Now we consider the <u>incidence graph</u> Γ of $\underline{S}$: the vertices of Γ are the points and blocks of $\underline{S}$, and two are joined if they are incident. This graph is bipartite, and vertices at distance i from a fixed vertex are joined only to vertices at distance i-1 and at distance i+1. Every vertex is joined to k others. Fixing, say, a point P in $\underline{S}$, we let D_i be the set of vertices at distance i from P, and $M_i = |D_i|$. Then

$M_0 = 1$, $M_1 = k$, $M_2 = k(k-1)/\lambda$ (see above). We shall prove

(a) $M_i \leq \frac{k(k-1)(k-\lambda)(k-\lambda-1)\ \ldots\ (k-\lambda-i+1)}{\lambda(\lambda+1)\ \ldots\ (\lambda+i-2)}$, $i \geq 2$.

(b) A vertex in D_i is joined to at least $\lambda+i-2$ vertices in D_{i-1}, $i \geq 1$.

Interpreting (a) appropriately, (a) and (b) are certainly true for $i = 2$. Suppose they are true for i, and suppose the elements in D_{i+1} are points, and let X be one of them. Then X is on at least one block $y \in D_i$, and by induction, y contains at least $\lambda+i-2$ points in D_{i-1}. Count flags (Y,z), where Y is one of the points of D_{i-1} on y, $z \neq y$ is a block on Y and on X. This number T is at least $(\lambda+i-2)(\lambda-1)$, since a point Y is on $\lambda-1$ blocks $z(\neq y)$ with X. On the other hand, if b is the number of such blocks z, then $T \leq b(\lambda-1)$, since a block z meets y in at most $\lambda-1$ more points (not all of which need lie in D_{i-1}). So $(\lambda+i-2)(\lambda-1) \leq b(\lambda-1)$, so $b \geq \lambda+i-2$ (since $\lambda > 1$). But the number of blocks of D_i on X is then at least $b+1$ (since y is not counted yet). So (b) is true for $i+1$.

Then counting flags (y,X), where $y \in D_i$, $X \in D_{i+1}$, (a) follows.

Now we see that Γ has bounded diameter, for $M_i > 0$ implies $k-\lambda-i+3>0$, or $i<k-\lambda+3$. So the number of vertices in Γ is $2v$, and

$$2v = \sum_{i=0}^{k-\lambda+2} M_i \quad .$$

More precisely:-

$$2v = 1 + k + \sum_{i=2}^{k-\lambda+2} M_i$$

and $M_i \leq \frac{(\lambda-1)!}{(k-2)\ldots(k-\lambda+1)} \binom{k}{\lambda+i-2}$ if $i \geq 2$, from (a).

Now

$$1 + k \leq \frac{(\lambda-1)!}{(k-2)\ldots(k-\lambda+1)} \left[\binom{k}{\lambda-2} + \binom{k}{\lambda-1}\right]$$

as can be seen by simplifying and using $\lambda \geq 2$.

But $\frac{(\lambda-1)!}{(k-2)\ldots(k-\lambda+1)} = \frac{\lambda-1}{\binom{k-2}{\lambda-2}}$

and so $2v \leq \frac{\lambda-1}{\binom{k-2}{\lambda-2}} \sum_{i=0}^{k-\lambda+2} \binom{k}{\lambda+i-2}$.

Thus $2v \leq \frac{\lambda-1}{\binom{k-2}{\lambda-2}} \sum_{j=\lambda-2}^{k} \binom{k}{j} \leq \frac{\lambda-1}{\binom{k-2}{\lambda-2}} \sum_{j=0}^{k} \binom{k}{j} = \frac{\lambda-1}{\binom{k-2}{\lambda-2}} 2^k$

so $v \leq (\lambda-1)2^{k-1}/\binom{k-2}{\lambda-2}$. ¤

The upper bound for v given in Theorem 21 is attainable when $\lambda = 2$, for every k. But it must be too large for $\lambda > 2$, and little is known about this situation. On the other hand, there is no upper bound in general when $\lambda = 1$:-

<u>Example 3</u> : Let $\underline{C}_1$ be a structure with one line and k points, all the points on the one line. We shall construct a series of structures

$$\underline{C}_1 \subset \underline{C}_2 \subset \ldots \subset \underline{C}_i \subset \ldots$$

such that in $\underline{C}_{2j}$ every point is on k lines and every line y in $\underline{C}_{2j}$ contains k points (if $y \in \underline{C}_{2j-1}$) or contains one point (if $y \notin \underline{C}_{2j-1}$), and dually for $\underline{C}_{2j+1}$, as follows :-

Consider $\underline{C}_{2j-1}$. There are points in $\underline{C}_{2j-1}$ on only one line, so we add k-1 new lines on each, every new line containing only one point. Then this new structure $\underline{C}_{2j}$ satisfies the condition above.

Now it is easy to prove inductively that the number of elements in $\underline{C}_i$ with only one incidence is a multiple of k. Suppose we consider $\underline{C}_n$, and suppose the elements with only one incidence are lines; partition these into subsets of k lines each, in any manner whatsoever. Suppose there is a projective plane $\underline{P}$ of order k-1 (e.g., k-1 is a prime power) and let P be a fixed point of $\underline{P}$. We identify the k lines through P with the k lines of one of the partitions in $\underline{C}_n$ above. Then we adjoin to each of these lines in $\underline{C}_n$ the remaining k-1 points of P on that line, and then the remaining $(k-1)^2$ lines of $\underline{P}$ are adjoined in the natural way to these points.

This yields a SSD with $\lambda = 1$, and clearly no upper bound exists for v. (The author conjectures that among the many obvious variations on the trick above, it is possible to choose one which works for any value of k, not just for "prime powers plus one".)

Now suppose $\underline{B}$ is a semi-biplane for (v,k,(2)), and let A be an incidence matrix for $\underline{B}$. Then

$$\begin{bmatrix} A & I \\ I & A^T \end{bmatrix}$$

is an incidence matrix of a semi-biplane for (2v,k+1,(2)), called the <u>double</u> of $\underline{B}$. Hence :-

<u>Lemma 22</u> : If there is a semi-biplane for (v,k,(2)), then there is one for (2v,k+1,(2)). In particular, there is a semi-biplane for

$(2^{k-1},k,(2))$, for every $k \geq 3$.

<u>Proof</u> : There is a trivial semi-biplane for $(4,3,(2))$ and its successive doubles (which are never trivial) give semi-biplanes for $(2^{k-1},k,(2))$. ¤

The upper limit for v in Theorem 21 is 2^{k-1} if $\lambda = 2$, so the semi-biplanes of Lemma 22 are maximal. Peter Wild has shown that they are also unique up to isomorphism (and can also be characterized by, e.g., their automorphism groups).

A SSD is <u>point divisible</u> if the relationship

$$P \sim Q \text{ if } P = Q \text{ or } P \text{ and } Q \text{ are not joined}$$

is an equivalence relation. With the obvious dual definition, we have :-

<u>Lemma 23</u> : A point divisible SSD is also block divisible.

<u>Proof</u> : Let y be a block of the point divisible SSD $\underline{S}$, with parameters $(v,k,(\lambda))$. No two points of y are equivalent, since they are on a common block (i.e., y). So if P is a point not on y, P can be equivalent to at most one point on y. Hence P is joined to $t = k$ or $k-1$ points on y.

We count flags (X,z), where X is on y and z is on P. The number is $t\lambda$, since there are t points X to choose from, so the number of z on P which meet y is also equal to t. Thus, in particular, P is on at most one block which does not meet y. So if w_1,w_2 are blocks not meeting y, they can have no point in common, and this proves that $\underline{S}$ is block divisible. ¤

In view of Lemma 23, we say that a SSD is <u>divisible</u> if it is point or block divisible. The number of points joined to a point P is $k(k-1)\lambda$, so the size of every divisibility class is $c = v-k(k-1)\lambda$. Clearly $m = v/c$ is an integer. But also:-

<u>Lemma 24</u> : If $\underline{S}$ is a divisible SSD for $(v,k,(\lambda))$, then $c \leq k/\lambda$ and $v \leq k^2/\lambda$.

<u>Proof</u> : Let $\underline{C}$ be a fixed divisibility class and P a point not in $\underline{C}$. A block on P contains at most one point of $\underline{C}$, so if s is the number of blocks on P which meet $\underline{C}$, and we count flags (X,z), where $X \in \underline{C}$, z on P, we find $s = c\lambda$, so $c = s/\lambda \leq k/\lambda$. Then $v \leq k(k-1)/\lambda + k/\lambda = k^2/\lambda$. ¤

An SSD with $v = k(k-1)/\lambda + 1$ is symmetric and trivially divisible with $c = 1$, while if $v = k(k-1)/\lambda + 2$, then it is divisible with $c = 2$, since every point determines a unique other point to which it is not joined. It is also easy to see that if $\underline{S}$ is a divisible SSD for $(v,k,(\lambda))$ with class size c, then there is a homomorphic image $\underline{S}'$ of $\underline{S}$, whose elements are the divisibility classes of $\underline{S}$, and $\underline{S}'$ is a symmetric design for $(v/c,k,\lambda c)$. This gives part (i) of the "Bose-Connor Theorem" for divisible SSD's :-

<u>Theorem 25</u> : If $\underline{S}$ is a divisible SSD for $(v,k,(\lambda))$ with class size c, then with $m = v/c$, we have :-

(a) there are integers x,y,z, not all zero, such that

$$x^2 = (k-\lambda c)y^2 + (-1)^{(m-1)/2}\lambda cz^2;$$

(b) there are integers x,y,z, not all zero, such that

$$x^2 = ky^2 + (-1)^{(v-m)/2} c^m z^2 .$$

The proof of this well-known theorem is similar to the proof of the Bruck-Ryser-Chowla Theorem (Theorem 7). The form of Theorem 25 is slightly different from the usual formulation of the Bose-Connor Theorem, but even in a more general form (for divisible PSD's) that theorem could be phrased in a similar way.

For $\lambda = 1$, the smallest divisible SSD whose existence is in doubt is (32,6,(1)), and for $\lambda = 2$, the smallest one is (38,9,(2)), as far as the author knows. It also may be worth noting that of the three semi-biplanes for (18,6,(2)), one is divisible (with c = 3) and the other two are not: so parameters alone do not in general imply divisibility.

Besides the generalised quadrangles and PSD's, there are many other families of 1-designs of interest and importance; most of them are generalizations of, or variationson, these two classes. We mention one additional family : the <u>semi-symmetric 3-designs</u>. While a 3-design can never be square unless it is trivial, a connected structure can satisfy : every three points (blocks) are on 0 or λ common blocks (points). These are always regular, uniform and square, and include the Higman-Sims geometry with 100 points and 22 points on a block, with $\lambda = 2$. (The Higman-Sims geometry is also an SSD for (100,22,(6)); indeed any semi-symmetric 3-design is an SSD.) There are infinitely many semi-symmetric 3-designs, but the other known ones are constructed by trickery from hadamard designs.

SOME REMARKS ON REPRESENTATION THEORY IN FINITE GEOMETRY

by

Udo Ott

University of Brunswick

Braunschweig, Germany

The purpose of this paper is to show in outline how the techniques and methods of ordinary representation and character theory are used in finite geometry.

The paper consists of five sections:

1. Hall's theorem and Baer's theorem
2. The spectrum of a graph
3. Foundations of geometry
4. The spectrum of a geometry
5. Representations and applications

Two of the most important sources concerning the development of subjects and methods in finite geometry are two theorems on finite projective planes due to Baer and Hall. In section one we present deep generalizations of these results.

In section two we sketch the classical method of using eigenvalues. In it we have tried to point out that the method of the k-spectrum leads to stronger results then the ordinary eigenvalue techniques over the reals. For example, we give an interesting proof of a theorem concerning the number of points in a strongly regular graph due to van Lint and Seidel.

In section three we consider basic topics of geometry: the definition, the flag graph, the open complex and the Steinberg module of a geometry.

Section four generalizes the methods mentioned in section two. We present a concept, which depends on structure theorems for non-commutative semisimple algebras. To illustrate the use of the various methods, we prove the most celebrated result concerning the order of a symmetric design due to Bruck, Ryser and Chowla.

In section five we give examples of representations and applications.

1. Hall's theorem and Baer's theorem

The Hall theorem on finite cyclic planes is the source for all research activity on quotient sets and multipliers. The theory of group algebras and their characters yields basic results on the existence of multipliers. We present a generalization of Bruck's theorem [8] :

Let G be a finite group and let $\lambda \geqslant 1$ be an integer. We say that a nonempty subset Δ of the group is a quotient set if every element $g \neq 1$ in G admits exactly λ representations of the form $g = xy^{-1}$ where x,y are elements in Δ. We note the obvious consequence:

(1) $\quad (|G|-1)\lambda = |\Delta|(|\Delta|-1)$.

An integer t is called a weak multiplier of the quotient set if $\Delta^t = \{x^t \mid x \in \Delta\} = \Delta g$ for suitable g in G.

With these preliminary remarks, we prove

Theorem 1.1 Let Δ be an invariant quotient set of the finite group G and let $p_1, p_2, \ldots, p_s$ be pairwise distinct prime divisors of $n = |\Delta| - \lambda = k - \lambda$, relatively prime to the exponent e of the group G, such that $p_1^{r_1} p_2^{r_2} \ldots p_s^{r_s} > \lambda$, where each r_i is the exponent of the prime p_i in the factorization of n. Then every integer $t \equiv p_i^{e_i} \pmod{e}$ for suitable $e_i \geqslant 1$, $i = 1, 2, \ldots, s$, is a weak multiplier.

The existence of weak multipliers rests also on the arithmetical structu[re] of the cyclotomic field of order e over $\mathbb{Q}$. So we introduce the numbe[r] field $K = \mathbb{Q}[\omega]$ where ω denotes a primitive eth root of unity over $\mathbb{Q}$.

One verifies easily that

(2) $\delta = \sum_{x \in \Delta} x \quad , \quad \bar{\delta} = \sum_{x \in \Delta} x^{-1} \quad , \quad \sigma = \sum_{x \in G} x$

are elements in the center of the group algebra $K[G]$. The assumption implies the equation

(3) $\delta \bar{\delta} = (k - \lambda) 1 + \lambda \sigma = n 1 + \lambda \sigma$.

If $\xi \neq 1_G$ is an irreducible character of the given group then the map $\frac{\xi}{\xi(1)}$ is a nonzero algebra homomorphism of the center of $K[G]$ onto K . It follows at once

(4) $\xi(\delta a) = \xi(\delta) \frac{\xi(a)}{\xi(1)} \quad , \quad a \in G \; ,$

and using (3) we deduce

(5) $\xi(\delta) \xi(\bar{\delta}) = n \, \xi(1)^2$.

As an almost immediate consequence of (4) and (5) , we obtain the <u>first basic Lemma</u> :

<u>Lemma 1.2</u> <u>Let</u> $\xi \neq 1_G$ <u>be an irreducible character of the group</u> G . <u>Then</u>

$$\xi(\delta a) \xi(\bar{\delta} b) = n \, \xi(a) \bar{\xi}(b)$$

<u>for all</u> $a, b \in G$.

We deduce a second basic Lemma: Let Γ be a map of a finite set M into the group G . The class function

(6) $\Phi_\Gamma(x) = |C_G(x)||\Gamma^{-1}(x^G)|$

can be expressed as a K-linear combination of irreducible characters ψ of G, say

$$\Phi_\Gamma = \sum_\psi c_\psi \psi , \qquad c_\psi \in K .$$

Let us denote the conjugacy classes in G by $K_1, \ldots, K_r$ and let g_i denote an element of class K_i. Then the orthogonality relations yield

$$c_\psi = (\Phi_\Gamma, \psi) = \sum_{x \in G} \frac{1}{|G|} \Phi_\Gamma(x) \bar{\psi}(x) =$$

$$= \sum_{x \in G} \frac{1}{|G|} |C_G(x)||\Gamma^{-1}(x^G)| \bar{\psi}(x) =$$

$$= \sum_{j=1}^{r} \frac{|K_j|}{|G|} |C_G(g_j)||\Gamma^{-1}(K_j)| \bar{\psi}(g_j) =$$

$$= \sum_{j=1}^{r} |\Gamma^{-1}(K_j)| \bar{\psi}(g_j) =$$

$$= \sum_{m \in M} \bar{\psi}(\Gamma(m)) = \bar{\psi}(\Gamma^*) ,$$

where $\Gamma^* = \sum_{m \in M} \Gamma(m)$. Therefore we obtain the decomposition

(7) $\Phi_\Gamma = \sum_\psi \bar{\psi}(\Gamma^*) \psi$.

We obtain an important special case for $M = \Delta a$, $a \in G$ and where $\Gamma : \Delta a \to G$ is the inclusion. Then, letting $\Phi_{\Delta a}$ be the corresponding class function, the equation (4) implies

(8) $\Phi_{\Delta a} = \sum_\psi \bar{\psi}(\delta) \dfrac{\bar{\psi}(a)}{\bar{\psi}(1)} \psi$.

Now we come to the <u>second basic Lemma</u> :

<u>Lemma 1.3</u> <u>Let</u> m <u>be an integer and let the map</u> $\Gamma : \Delta \to G$ <u>be defined by the rule</u> $\Gamma(x) = x^m$. <u>Then</u>

$$|G||\Gamma^{-1}(\Delta g)| = |G||\{x \in \Delta \mid x^m \in \Delta g\}| =$$

$$= k^2 + \sum_{\xi \neq 1_G} \frac{\xi(g)}{\xi(1)} \xi(\delta) \overline{\xi}(\delta^*) ,$$

where $\delta^* = \sum_{x \in \Delta} x^m$.

<u>Proof.</u> Using (7) and (8) we compute easily

$$(\Phi_\Gamma, \Phi_{\Delta g}) = \sum_\psi \frac{\psi(g)}{\psi(1)} \psi(\delta) \overline{\psi}(\Gamma^*) =$$

$$= k^2 + \sum_{\xi \neq 1_G} \frac{\xi(g)}{\xi(1)} \xi(\delta) \overline{\xi}(\delta^*) .$$

We denote the conjugacy classes in such a way that $\Delta^m = \bigcup_{j=1}^{t} K_j$ and establish that the integer $|\Gamma^{-1}(x)| = f_j$ is independent of $x \in K_j$. Therefore we have

$$(\Phi_\Gamma, \Phi_{\Delta g}) = \frac{1}{|G|} \sum_{x \in G} \Phi_\Gamma(x) \overline{\Phi_{\Delta g}(x)} =$$

$$= \frac{1}{|G|} \sum_{j=1}^{r} |K_j||C_G(g_j)|^2 |\Delta g \cap K_j||\Gamma^{-1}(K_j)| =$$

$$= \sum_{j=1}^{t} \frac{1}{|G|} |K_j|^2 |C_G(g_j)|^2 f_j |\Delta g \cap K_j|$$

$$= \sum_{j=1}^{t} |G||\Delta g \cap K_j| f_j = |G||\Gamma^{-1}(\Delta g)| ,$$

and the Lemma is proved. □

We are now ready to prove the main theorem 1.1. For $1 \leq i \leq s$, let $p = p_i$, $f = e_i$, $r = r_i$ and $q = p^f$. If σ denotes the Frobenius automorphism of the number field K, determined by the prime p, then for all irreducible characters of G we have $\xi(g^q) = \xi(g)^\tau$, where $\tau = \sigma^f$. Now the second basic Lemma yields

$$(9) \qquad |G||\Delta g \cap \Delta^q| = k^2 + \sum_{\xi \neq 1_G} \frac{\xi(g)}{\xi(1)} \xi(\delta) \overline{\xi}(\delta)^\tau .$$

A remarkable property of the Frobenius automorphism is the fact that σ leaves a prime ideal P in K above p fixed. Let υ be the exponent of K given by the prime ideal P. By assumption, the prime p does not divide the exponent of G, hence p is not ramified, and consequently $\upsilon(n) = r$. By (9) we have

$$\upsilon(|G||\Delta g \cap \Delta^q| - k^2) =$$

$$= \upsilon\Big(\sum_{\xi \neq 1_G} \frac{\xi(g)}{\xi(1)} \xi(\delta) \overline{\xi}(\delta)^\tau\Big) \geq$$

$$\geq \min_{\xi \neq 1_G} \upsilon\Big(\frac{\xi(g)}{\xi(1)} \xi(\delta) \overline{\xi}(\delta)^\tau\Big) .$$

Since $\upsilon(\xi(g)) = \upsilon(\xi(1)) = 0$ it follows that

$$\upsilon\Big(\frac{\xi(g)}{\xi(1)} \xi(\delta) \overline{\xi}(\delta)^\tau\Big) = \upsilon(\xi(\delta) \overline{\xi}(\delta)^\tau) =$$

$$= \upsilon(\xi(\delta) \overline{\xi}(\delta)) = \upsilon(n \xi(1)^2) = r ,$$

and this gives

(10) $\quad |G|\,|\Delta g \cap \Delta^q| \equiv k^2 \pmod{p^r}$.

Then by (1) , we have

(11) $\quad |\Delta g \cap \Delta^q| \equiv \lambda \pmod{a}$,

where $\quad a = p_1^{r_1} \dots p_s^{r_s}$.

We begin the final step of the proof with the observation that since $|\Delta g \cap \Delta^t| \equiv \lambda \pmod{a}$, the assumption implies that $|\Delta g \cap \Delta^t| = \lambda(g) \geqslant \lambda$. We consider an element $h \notin \Delta^t$. One computes

$$\sum_{h \in \Delta g} \lambda(g) = |\Delta^t|\,\lambda \quad ,$$

hence

$$\sum_{h \in \Delta g} (\lambda(g) - \lambda) = 0 \quad ,$$

and therefore we conclude $\lambda(g) = \lambda$ provided that there is an element $h \in \Delta g \setminus \Delta^t$. Assume now $\Delta^t \neq \Delta g$ for all $g \in G$. It follows $\lambda(g) = \lambda$. Considering an element $h \in \Delta^t$ we obtain

$$(|\Delta^t| - 1)\,\lambda = \sum_{h \in \Delta g} (\lambda(g) - 1) = (\lambda - 1)\,k \quad ,$$

hence $\lambda = k$. But then $G = \Delta = \Delta^t$, a contradiction. The theorem is proved. □

Now we turn to the second main result on finite projective planes, that is Baer's theorem [1] . Baer's general idea of counting certain elements in the Steinberg module of the projective plane (to be defined in section 4) has been lost because there is a particularly short proof of his result resting on the method of "incidence matrices". Baer's theorem and the just mentioned method have supported a great number of results on the influence of correlations on the structure of finite geometries. We shall give information about generalized polygons in sections 2 and 4 after having introduced suitable methods. Here we present a deep generalization of Baer's result using the theory of semi-simple rings. Strictly speaking it is a generalization of theorems proved by Hoffman, Newman, Straus, Taussky [21] and Ball [2]:

Theorem 1.4 Let γ be a correlation of a design, with parameters v, $k = 1+q$, λ. Suppose that not every generator of the group $H = \langle\gamma\rangle$ admits exactly k isotropic points. Then the order $n = k - \lambda$ is a square if and only if all generators have the same number of isotropic points. If this number is denoted by $\mu + 1$ then we have $\mu \equiv q \pmod{\sqrt{n}}$.
If the order of the design is not a square then there are two different integers α and β such that half of the generators have exactly $\alpha + 1$ and the other half have exactly $\beta + 1$ isotropic points. Write the order as $n = t^2 s$, where $s \neq 1$ is the square-free part of n. Then $\alpha + \beta = 2q$ and $\alpha, \beta \equiv q \pmod{ts}$. Furthermore, the integer s divides $\frac{1}{2} o(\gamma)$, even $\frac{1}{4} o(\gamma)$ if $o(\gamma) \equiv 0 \pmod 4$.

The key point in the proof is the fact that a certain algebra A is semi-simple. The algebra is defined over the field $K = \mathbb{Q}\sqrt{n}$ with generators Γ and Λ, and relations

$$\Gamma^2 = q1 + (q-1)\Gamma$$

$$(12) \qquad \Lambda^{2m} = 1, \quad \Lambda^2\Gamma = \Gamma\Lambda^2$$

$$(\lambda-1)\Gamma + \Gamma\Xi\Gamma = (\lambda-1)\Xi + \Xi\Gamma\Xi, \quad \Xi = \Lambda^{-1}\Gamma\Lambda$$

where $2m > 2$ denotes the order of the given correlation.

Let ω denote a $2m$-th primitive root of unity and set $L = K[\omega]$. As usual the algebra A is embedded in $B = A \otimes_K L$ by means of the isomorphism defined by $x \mapsto x \otimes 1$, $x \in A$. The proof depends on the following result.

Lemma 1.5 *Let δ denote a $2m$-th root of unity. Then the L-algebra B admits one-dimensional representations ind_δ and st_δ with*

$$ind_\delta(\Gamma) = q \quad , \quad ind_\delta(\Lambda) = \delta$$

$$st_\delta(\Gamma) = -1 \quad , \quad st_\delta(\Lambda) = \delta$$

and a two-dimensional irreducible representation $\mathfrak{F}_\delta$ with

$$\mathfrak{F}_\delta(\Gamma) = \begin{pmatrix} -1 & \sqrt{n} \\ 0 & q \end{pmatrix} , \quad \mathfrak{F}_\delta(\Lambda) = \begin{pmatrix} 0 & \delta \\ \delta & 0 \end{pmatrix} .$$

If χ_δ denotes the character of this two-dimensional representation then we have $\chi_\delta(\Lambda^i\Gamma) = \delta^i\sqrt{n}$ *if* $i \equiv 1 \pmod 2$.

Proof. A simple verification shows that with the substitution

$$\Gamma \longleftrightarrow q \quad , \quad \Lambda \longleftrightarrow \delta \; ,$$

respectively

$$\Gamma \longleftrightarrow \begin{pmatrix} -1 & \sqrt{n} \\ 0 & q \end{pmatrix} , \quad \Lambda \longleftrightarrow \begin{pmatrix} 0 & \delta \\ \delta & 0 \end{pmatrix}$$

the relations (12) will be verified. □

As an immediate consequence, we obtain that $\dim_K A = \dim_L B \geq 12m$.
We infer from the relations (12) that
$\{\Lambda^i, \Lambda^i\Gamma, \Lambda^i\Xi, \Lambda^i\Gamma\Xi, \Lambda^i\Xi\Gamma, \Lambda^i\Gamma\Xi\Gamma \mid 1 \leq i \leq 2m\}$ is a set of generators for the algebra A . We conclude that $\dim_K A = \dim_L B = 12m$.
This last result has the important consequence: The algebra B is semisimple and each irreducible B - module is isomorphic to a module mentioned in Lemma 1.5.

Now we turn to the geometry. We begin with the presentation of an A - module. Let F be the set of all flags of the design. The K - space V of functions $f : F \to K$ becomes a left A - module, if we define

$$(\Gamma f)\{A,a\} = \sum_{\substack{X\,I\,a \\ X \neq A}} \{X,a\} \qquad \{A,a\} \in F .$$

$$(\Lambda f)\{A,a\} = \{A^\gamma, a^\gamma\}$$

Indeed, one can easily verify that the relations (12) are fullfilled for the linear maps $f \mapsto \Gamma f$ and $f \mapsto \Lambda f$.
We consider the B - module $W = V \otimes_K L$. The subalgebra of B generated by the two elements Γ and Ξ admits two important B - invariant submodules: $W_o = \{f \mid \Gamma f = \Xi f = qf\}$ and $W_1 = \{f \mid \Gamma f = \Xi f = -f\}$.
It is easily seen that $\dim_L W_o = 1$. Furthermore, the B - module W_o is isomorphic to the one-dimensional representation ind_1 mentioned in Lemma 1 .5.
If we denote the character of W resp. W_1 by Φ resp. χ , then we obtain the decomposition

$$(13) \qquad \Phi = \mathrm{ind}_1 + \chi + \sum_{\delta} n_\delta \chi_\delta \quad ,$$

where n_δ is the multiplicity of the irreducible component of W associated with $\mathfrak{F}_\delta$ in the direct decomposition of W into irreducible submodules.

Now there are two possibilities to complete the proof. One possibility rests on the notion of algebraic conjugate representations; since the B-module W is defined over the field K algebraic conjugate representations appear with the same multiplicity. But we take the second way: If the correlation γ leaves exactly $f+1$ flags fixed then we have for all integers j relatively prime to $2m$ the equation $\Phi(\Lambda^j) = 1 + f$. Now, using the fact $\chi_\delta(\Lambda^j) = 0$, we obtain $\chi(\Lambda^j) = f$, hence $\chi(\Lambda^j\Gamma) = -\chi(\Lambda^j) = -f$. Because of equation (13), one obtains

$$(14) \qquad \Phi(\Lambda^j\Gamma) + f - q = \sum_\delta n_\delta \delta^j \sqrt{n} .$$

It follows at once from the definition that the correlation γ^j admits exactly $\Phi(\Lambda^j\Gamma) + f + 1 = \alpha_j + 1$ isotropic points, and we obtain

$$(15) \qquad \alpha_j - q = \sqrt{n} \sum_\delta n_\delta \delta^j .$$

We shall now show that $\alpha_j = \alpha_1$, if the order n is a square. The cyclotomic field $\mathbb{Q}[\omega]$ allows a Galois automorphism which maps δ to δ^j. Applied to the equation $\alpha_1 - q = \sqrt{n} \sum_\delta n_\delta \delta$ this automorphism yields $\alpha_1 - q = \alpha_j - q$, as asserted. Conversely, suppose $\alpha_1 = \alpha_j$ for every integer j relatively prime to $2m$. From (15) we have

$$\varphi(2m)(\alpha_1 - q) = \sqrt{n} \sum_\delta n_\delta \sum_{\substack{1 \leq j \leq 2m \\ (j,2m)=1}} \delta^j .$$

By the well-known fact that $\sum_{\substack{1 \leq j \leq 2m \\ (j,2m)=1}} \delta^j$ is an integer, we conclude $\sqrt{n} \in \mathbb{Q}$. Thus n is a square and $\alpha_1 \equiv q \pmod{\sqrt{n}}$.

Suppose now that the order n is not a square, say $n = t^2 s$ with square-free part $s \neq 1$. Then equation (15) implies that $\sqrt{n} \in \mathbb{Q}[\omega]$ and we conclude already that s divides m, even $\frac{1}{2}m$ in the case $m \equiv 0 \pmod 2$. The Galois group of $\mathbb{Q}[\omega]$ over $\mathbb{Q}$ possesses a subgroup of index two admitting the fixed field $\mathbb{Q}[\sqrt{n}]$. Since a Galois automorphism maps $\sqrt{n}$ to $\pm\sqrt{n}$, equation (15) yields the remaining statements. The theorem is proved. □

2. The spectrum of a graph

A number of problems about the structure of finite geometries are intimately related to the study of symmetric relations. For example, a polarity π of a finite design defines a symmetric irreflexive relation on the set of points by the rule

(1) $A \sim B$, if A and B^{π} are incident and $A \neq B$.

In this section we give a simple, but important, method to study such relations, which depends upon structure theorems for semi-simple commutative algebras. In section 4 we shall present a general method, which also depends on structure theorems for non-commutative algebras.

In the following we prefer the terminology of graph theory. A graph G is a symmetric irreflexive (binary) relation on a vertex set Ω . If $\{A,B\}$ is an edge of G , then the vertices A and B are said to be adjacent, written $A \sim B$.

Let k be a field of characteristic zero. The set $V = M(\Omega,k) = V_k$ of all maps of Ω into k forms a vector space over the field k . We find at once

(2) $\dim_k V = |\Omega|$.

The vector space V_k carries the structure of an orthogonal space, if we define a symmetric bilinear form by the rule

(3) $(f,g) = \sum_{A \in \Omega} f(A)\, g(A)$.

The important adjacency map $\alpha : V \to V$ is defined by

(4) $$(\alpha f)(A) = \sum_{X \sim A} f(X) \quad .$$

One easily checks that α is a linear self-adjoint map of the orthogonal space. We obtain the adjacency algebra $A = k[\alpha]$ of the graph by substituting α for a transcendental t in polynomials in t over k. Clearly, the vector space V_k becomes a faithful A-module. We define the standard basis $\{ \chi_A | A \in \Omega \}$ of the standard module V by means of

(5) $$\chi_A(B) = \delta_{AB} \quad .$$

The adjacency matrix of the graph is the symmetric matrix of α with respect to the standard basis. Because the field k has characteristic zero and since the self-adjoint map α is defined over the field $\mathbb{Q}$ of rational numbers (clearly, we have $A_k = A_{\mathbb{Q}} \otimes_{\mathbb{Q}} k$ and $V_k = V_{\mathbb{Q}} \otimes_{\mathbb{Q}} k$), the adjacency algebra is semi-simple.

The k-spectrum of the graph is a full set $\{L_1, L_2, \ldots, L_r\}$ of non-isomorphic irreducible A-modules, together with their multiplicities $n_1, n_2, \ldots, n_r \geq 1$ in a direct decomposition of the standard module into irreducible submodules. We write

(6) $$\mathrm{spec}_k(G) = \begin{pmatrix} L_1 & L_2 & \ldots & L_r \\ n_1 & n_2 & \ldots & n_r \end{pmatrix} \quad .$$

Thus if χ_j is the character afforded by L_j we may write

(7) $$\Phi = \sum_{j=1}^{r} n_j \chi_j \quad ,$$

where Φ denotes the standard character of the standard module.

Another important invariant which will prove to be useful is the discriminant of an orthogonal space. Corresponding to an orthogonal decomposition of the standard module

(8) $V = V_1 \perp \ldots \perp V_t$,

we have

(9) $\prod_{j=1}^{t} d_j$ is a square in k ,

where d_j denotes a discriminant of the submodule V_j .

We see now that our study of symmetric relations induces three main problems: The determination of all irreducible A-modules, the determination of their multiplicities in the standard module, the determination of the various discriminants.

If the minimal polynomial of α has all zeros in k , then a full set of non-isomorphic irreducible A-modules can be characterized in terms of the eigenvalues $e_1, e_2, \ldots, e_r$ of α as follows: there are exactly r non-isomorphic one-dimensional A-modules, say $L_1, L_2, \ldots, L_r$, the action of α on L_j being given by multiplication with e_j. Thus if χ_j is the character afforded by L_j we have $\chi_j(\alpha) = e_j$.

In connection with (7) the following geometrical interpretation of some character values is of practical importance for computations.

<u>Theorem 2.1</u> <u>The number of closed walks of length</u> m <u>is equal to</u> $\Phi(\alpha^m)$.

As an illustration, we present the foundations of the theory of <u>strongly</u>

<u>regular graphs</u> (as a general reference for this theory, we may list Seidel [31] and Cameron [9]) : The graph G is <u>strongly regular</u>, if G is regular of valency ζ, and if two distinct vertices A and B are adjacent to exactly λ respectively μ vertices if $A \sim B$ respectively $A \nsim B$.

Suppose that G is a regular graph of valency ζ. Then the function ι, given by

$$(10) \qquad \iota(A) = 1 \quad , \quad A \in \Omega \quad ,$$

is an eigenvector of α with eigenvalue ζ. If the graph is connected, this eigenvalue is of multiplicity one (of course the multiplicity over k equals the multiplicity over the rational field). Since the definition of a strongly regular graph implies that α is a zero of a polynomial of degree three, we have apart from the eigenvalue ζ of multiplicity one or two additional eigenvalues. By theorem 2.1 and (7) we find their multiplicities:

<u>Theorem 2.2</u> <u>Let G be a strongly regular graph with the vertex-set Ω and the parameters</u> $n = |\Omega|, \zeta, \lambda$ <u>and</u> $\mu \geqslant 1$.

(<u>Type</u> A) <u>If</u> $(n-1)(\mu-\lambda) \neq 2\zeta$, <u>then the rational field is a splitting field for the adjacency algebra. We have</u>

$$\mathrm{spec}_{\mathbb{Q}}(G) = \begin{pmatrix} \zeta & r^+ & r^- \\ 1 & f^+ & f^- \end{pmatrix} ,$$

<u>where</u>

$$r^{\pm} = \frac{1}{2}\left(\lambda - \mu \pm \sqrt{4(\zeta-\mu) + (\lambda-\mu)^2}\right)$$

$$f^{\pm} = \frac{1}{2}\left(n - 1 \pm \frac{(n-1)(\mu-\lambda) - 2\zeta}{\sqrt{4(\zeta-\mu) + (\lambda-\mu)^2}}\right) \quad .$$

(Type B) *If* $(n-1)(\mu-\lambda) = 2\zeta$ *, then* $k = \mathbb{Q}[\sqrt{n}]$ *is a splitting field for the adjacency map. We have*

$$\mathrm{spec}_k\,(G) = \begin{pmatrix} \zeta & r^+ & r^- \\ 1 & \frac{n-1}{2} & \frac{n-1}{2} \end{pmatrix} ,$$

where $r^{\pm} = \frac{1}{2}(-1 \pm \sqrt{n})$.

In case B it may very well happen that the number n of vertices is not a square. In any case, however, there is a well-known condition due to van Lint and Seidel:

Theorem 2.3 *If* G *is a strongly regular graph of type* B *, then the number* n *of vertices is a sum of two squares of integers.*

Proof. Suppose that n is not a square. By the preceding theorem we know that $k = \mathbb{Q}[\sqrt{n}]$ is a splitting field for the adjacency algebra, and that the standard module allows an orthogonal decomposition

(11) $$V_k = V_1 \perp V^+ \perp V^- ,$$

where $V_1 = \langle \iota \rangle$ and $V^{\pm}$ is the eigenspace of α corresponding to the eigenvalue $r^{\pm}$. Let $- : x \mapsto \bar{x}$ denote the Galois automorphism of k . We apply the concept of looking down to the rational field using the semilinear map $\wedge : V_k \to V_k$, which is defined by means of

(12) $$(\wedge f)(A) = \overline{f(A)} .$$

Because $\wedge\alpha = \alpha\wedge$, $(\wedge f, \wedge g) = \overline{(f,g)}$ and $\overline{r^+} = r^-$, we have

$$\wedge(V^+) = V^-$$
$$\wedge(V^-) = V^+ .$$

Moreover, we obtain a decomposition of the rational standard module

(13) $V_{\mathbb{Q}} = C_{V_1}(\wedge) \perp C_W(\wedge)$,

where $W = V^+ \oplus V^-$.

Now let $\{f_1, \ldots, f_{\frac{n-1}{2}}\}$ be an orthogonal basis of V^+ . It follows that $\{\wedge f_1, \ldots, \wedge f_{\frac{n-1}{2}}\}$ is an orthogonal basis of V^- . The submodules $W_i = \langle f_i, \wedge f_i \rangle$ are $\wedge$-invariant, and we have

$$W = W_1 \perp W_2 \perp \ldots \perp W_{\frac{n-1}{2}} .$$

We obtain

$$C_W(\wedge) = C_{W_1}(\wedge) \perp \ldots \perp C_{W_{\frac{n-1}{2}}}(\wedge) .$$

It is easy to calculate the discriminant of $C_{W_i}(\wedge)$: The submodule is generated by the two functions $f_i + \wedge f_i$ and $\sqrt{n}\, f_i - \sqrt{n}\, \wedge f_i$. Because $(f_i, \wedge f_i) = 0$, it follows that $4n\,(f_i, f_i)\overline{(f_i, f_i)} =$ $= 4n \operatorname{Norm}((f_i, f_i))$ is the discriminant. Trivially, $(\iota, \iota) = n$.

Using (13) we conclude that $n\, 4^{\frac{n-1}{2}}\, n^{\frac{n-1}{2}} \operatorname{Norm}\left(\prod_{i=1}^{\frac{n-1}{2}} (f_i, f_i)\right)$ is a square in $\mathbb{Q}$.

It is clear that $\frac{n-1}{2} \equiv 0 \pmod 2$. Then it is immediate that $n\,N(z) = n\,x^2 - y^2$ is a square in $\mathbb{Q}$ for a suitable $z = x + y_o\sqrt{n} \in k$, $x, y_o, y \in \mathbb{Q}$. Thus we have proved that n is a sum of two squares of rational numbers. By a standard argument, n is a sum of two squares of integers. The theorem is proved. □

Another interesting application yields a result on polarities of generalized quadrangles, which was proved by Payne [28] and Thas [34] using the symmetric and irreflexive relation mentioned in the beginning of this section:

Theorem 2.4 *If a generalized quadrangle of order q allows a polarity, then the integer $2q$ is a square.*

The results described in the last statements are based on the fact that the dimension of the adjacency algebra is small. In the more general case of arbitrary dimension, usually the calculation of multiplicities rests on orthogonality relations. The adjacency algebra A is a symmetric algebra with respect to the bilinear form $(\ ,\ ) : A \times A \rightarrow k$ defined by the formula

$$(14) \qquad (x,y) = \frac{1}{|\Omega|}\,\Phi(xy) \quad , \quad x,y \in A \quad .$$

This bilinear form is obviously associative and symmetric. Because the adjacency algebra is semi-simple, it follows easily that the form is non-degenerate. Let $\{\alpha_1 = 1, \alpha_2, \ldots, \alpha_s\}$ and $\{\beta_1 = 1, \beta_2, \ldots, \beta_s\}$ be dual bases of the adjacency algebra A. As is customary, with respect to the dual bases we introduce a bilinear form on the dual space A^* by the rule

$$(15) \qquad [\psi,\chi] = \sum_{j=1}^{s} \psi(\alpha_j)\,\chi(\beta_j) \quad , \quad \psi,\chi \in A^* \quad .$$

The importance of this bilinear form rests on

Lemma 2.5 *Let ψ, χ be the characters of two irreducible non-isomorphic A-modules. Then $[\psi,\chi] = 0$.*

In terms of the earlier notation of the k-spectrum of a graph, we may write the standard character $\Phi = \sum_{j=1}^{s} n_j \chi_j$. We evidently have $[\Phi, \chi_i] = |\Omega|\, \chi_i(1)$; but on the other hand the orthogonality relations yield $[\Phi, \chi_i] = n_i [\chi_i, \chi_i]$.

Therefore we obtain the fundamental formula

(16) $$n_i = \frac{|\Omega|\, \chi_i(1)}{[\chi_i, \chi_i]} , \qquad 1 \leqslant i \leqslant s .$$

As an application of the preceding statements we mention the celebrated structure theorem on Moore graphs (Bannai and Ito [3], Damerell [11], compare with Biggs [4]): A graph G of valency $\zeta > 2$ and diameter $d \geqslant 2$ is called a <u>Moore graph</u> if each pair of vertices of the graph is joined by a unique path (walk without repeated vertices) of length at most d .

If Ω is the vertex-set of the Moore graph, we partition $\Omega \times \Omega \setminus \{(A,A) \mid A \in \Omega\}$ into graphs $\underset{i}{\sim}$, $1 \leqslant i \leqslant d$, where $\underset{i}{\sim}$ is given by

(17) $A \underset{i}{\sim} B$, if there is a path of length i joining the vertices A and B .

Let $\alpha = \alpha_1, \alpha_2, \ldots, \alpha_d$ be the set of the corresponding adjacency maps. An easy induction argument yields

(18) $$\alpha\alpha_i = (\zeta - 1)\alpha_{i-1} + \alpha_{i+1} , \qquad 2 \leqslant i \leqslant d-1 ,$$

showing that the adjacency algebra $A = k[\alpha]$ contains all adjacency maps. Moreover, the adjacency maps are linearly independent over k , and because $\Phi(\alpha_i \alpha_j) = 0$ for $i \neq j$ we obtain an orthogonal basis $\{\alpha_o = 1, \alpha_1, \ldots, \alpha_d\}$ for A. Thus we have found a natural pair of dual bases for the adjacency algebra, suitable for character calculations. The missing structure equation

$$(18^*) \qquad \alpha\alpha_d = (\zeta - 1)(\alpha_{d-1} + \alpha_d)$$

implies that, apart from the eigenvalue ζ, the eigenvalues of α are

$$e_i = 2\sqrt{\zeta - 1}\cos\varphi_i \quad , \qquad 1 \leqslant i \leqslant d \quad ,$$

for suitable $0 < \varphi_i < \pi$. This representation of the eigenvalues and the fact that by (17)

$$[\chi_i, \chi_i] = \sum_{j=o}^{d} \frac{1}{(\alpha_j, \alpha_j)} \chi_i(\alpha_j)^2$$

is a rational number makes it fairly clear that there are only few eigenvalues. However, we emphasize that this judgment requires extensive character calculations.

The major result then is

Theorem 2.6 *There are only Moore graphs of diameter* $d = 2$.

3. Foundations of geometry

In the preceding section, we established a method to study symmetric relations, which depends upon structure theorems for semi - simple commutative rings. We shall present a powerful generalization of the method to the case where the graph admits an edge - coloring. But we shall be primarily interested in the flag graph of a geometry. Therefore it seems desirable to first consider basic topics of (finite) geometry and to present the generalization in the next section.

For additional information, see Buekenhout [6] and Tits [35].

We start with a basic definition:

A geometry $G = (\Omega_1, \ldots, \Omega_n ; J)$ of rank n is a collection of pairwise disjoint sets $\Omega_1, \ldots, \Omega_n$ and a symmetric reflexive relation J, which is defined on the basic set $\Omega = \bigcup_{j=1}^{n} \Omega_j$ such that

(G 1) For $\omega, \tau \in \Omega_i$, the relation $\omega J \tau$ implies that $\omega = \tau$,

(G 2) A maximal flag contains exactly n elements.

A subset F of Ω is called a flag if J restricted to F is trivial:

(1) $F \subseteq \Omega$ is a flag, if $\omega J \tau$ for all $\omega, \tau \in F$.

The binary relation J is called the incidence relation of the geometry. It is also convenient to introduce the following term: We say that an element $\omega \in \Omega$ is of type i provided ω is in Ω_i.

Perhaps this is the most important definition for a geometry. A similar idea of a (linear) geometry was first discovered by Steinitz [32], [33]. The keystone of Steinitz's theory of polyhedral geometries is the

following definition of the flag graph of a geometry:
The vertex-set of the flag graph is the set F of all maximal flags of the geometry G. Two maximal flags F_1 and F_2 are adjacent (resp. i-adjacent) if they differ in exactly one element (of type i); we then write $F_1 \sim F_2$ (resp. $F_1 \underset{i}{\sim} F_2$). For each type i we obtain an equivalence relation $\underset{i}{\equiv}$ which is defined by the rule

(2) $F_1 \underset{i}{\equiv} F_2$ if $F_1 = F_2$ or $F_1 \underset{i}{\sim} F_2$.

We call the geometry regular if each class of the equivalence relation $\underset{i}{\equiv}$ contains $1 + q_i \geq 2$ elements.

The next concept to be introduced is that of cohomology, which is used to define the Steinberg module of the geometry. In order to construct the complex of G, we shall give some further notation and definitions. For our purposes in this section, we shall take a somewhat more special point of view: We let k denote a field of characteristic zero. For m a positive integer, let F_m denote the set of all flags of cardinality m. Then the set $V^m = M(F_m, k)$ of all maps of F_m into k forms a vector space over the field k. We have $(o) = V^o = V^{n+1} = V^{n+2} = \ldots$. For convenience we write V for V^n. The set $T = \{1,2,\ldots,n\}$ of types induces an order on each flag. We define the coboundary map $d\colon V^m \to V^{m+1}$ by

(3) $$(df)(H) = \sum_{j=1}^{m+1} (-1)^{j-1} f(H_j) .$$

Here the notation H_j means omit ω_{i_j} in the ordered set $H = \{\omega_{i_1} < \omega_{i_2} < \ldots < \omega_{i_{m+1}}\}$. One checks easily that $d^2 = 0$; thus we have

Lemma 3.1 *The sequence* $0 \to 0 \to V^1 \xrightarrow{d} V^2 \xrightarrow{d} \ldots \xrightarrow{d} V^n \to 0 \to 0$ *is an (open) complex of vector spaces.*

The *Steinberg module* of the geometry is the n-th cohomology group $H_n = V/_{d(V^{n-1})}$ of this complex. The Euler - Poincaré characteristic with respect to the dimension of spaces can be used for calculating the dimension of the Steinberg module in some easy cases. Results dealing with this problem for particular geometries may be found in Ronan [29]. We conclude this section with an easy consequence of the definitions:

Lemma 3.2 *Let* $G = (\Omega_1, \Omega_2 ; J)$ *be a geometry of rank* $n = 2$. *Then the Steinberg module of* G *is of dimension*

$$|F| - |\Omega_1| - |\Omega_2| + 1 .$$

4. The spectrum of a geometry

Many problems concerning finite geometries, particularly questions about their arithmetical invariants or questions about their automorphisms, can be transformed into questions concerning the associated Hecke algebra (or incidence algebra). We now describe this algebra and begin with a few definitions, which generalize the situation of section 2.

Let G, n, F, k and $V = V_k$ have the same meanings as in the preceding section. As we have seen in section 2, (3), the standard module V of the geometry carries the structure of an orthogonal space:

(1) $$(f,g) = \sum_{H \in F} f(H)\, g(H) \quad , \qquad f,g \in V \ .$$

The important adjacency map $\sigma_i : V \to V$ of type i is defined by

(2) $$(\sigma_i f)(H) = \sum_{G \underset{i}{\sim} H} f(G) \ .$$

The linear map σ_i is a self-adjoint map of the orthogonal space. This follows readily from the fact that the linear map is represented by a symmetric matrix with respect to the standard basis $\{ \chi_H \mid H \in F \}$.

Definition 4.1 The Hecke algebra $H_k(G) = H$ (or incidence algebra) is defined to be the subalgebra of $\mathrm{End}_k(V)$ given by all polynomials in $\sigma_1, \sigma_2, \ldots, \sigma_n$.

Clearly, the standard module becomes a faithful H-module. The passage to the adjoint of a linear map with respect to our form induces an antiautomorphism:

Lemma 4.2 The Hecke algebra $H_k(G)$ has an antiautomorphism which fixes $\sigma_1, \sigma_2, \ldots, \sigma_n$.

Because the field has characteristic zero and since the Hecke algebra is defined over the field of rational numbers ($H_k(G) = H_{\mathbb{Q}}(G) \otimes_{\mathbb{Q}} k$, $V_k = V_{\mathbb{Q}} \otimes_{\mathbb{Q}} k$) , lemma 4.2 yields

Theorem 4.3 The Hecke algebra is semi - simple.

The k - spectrum of the geometry is a full set $L_1, L_2, \ldots, L_r$ of nonisomorphic irreducible H - modules, together with their multiplicities $n_1, n_2, \ldots, n_r \geq 1$ in a direct decomposition of the standard module into irreducible submodules. We write

$$(3) \qquad \operatorname{spec}_k(G) = \begin{pmatrix} L_1 & L_2 & \ldots & L_r \\ n_1 & n_2 & \ldots & n_r \end{pmatrix} ,$$

or

$$\operatorname{spec}_k(G) = \begin{pmatrix} \chi_1 & \chi_2 & \ldots & \chi_r \\ n_1 & n_2 & \ldots & n_r \end{pmatrix} ,$$

where χ_j is the character afforded by L_j . Thus if Φ is the standard character afforded by the standard module we may write

$$(4) \qquad \Phi = \sum_{j=1}^{r} n_j \chi_j .$$

As in the case of adjacency algebras, the study of a finite geometry leads to three major themes:

(I) Determination of irreducible H - modules

(II) Determination of multiplicities

(III) Determination of discriminants .

As an illustration, we present the spectrum of a symmetric design mentioned in the proof of theorem 1.4:

Theorem 4.3 Let $G = (\Omega_1, \Omega_2 ; J)$ be a symmetric design, with parameters $v, k = 1+q, \lambda$. Then $\dim_k H_k(G) = 6$ and $H_k(G)$ has defining relations

$$\sigma_i^2 = q\cdot 1 + (q-1)\,\sigma_i \quad , \quad i = 1,2$$

$$(\lambda - 1)\,\sigma_1 + \sigma_1\sigma_2\sigma_1 = (\lambda - 1)\,\sigma_2 + \sigma_2\sigma_1\sigma_2 \quad .$$

The Hecke algebra admits one - dimensional representations ind and st with

$$\text{ind}\,(\sigma_1) = \text{ind}\,(\sigma_2) = q$$

$$\text{st}\,(\sigma_1) = \text{st}\,(\sigma_2) = -1 \quad ,$$

and a two - dimensional irreducible representation $\mathfrak{F}$ with

$$\mathfrak{F}(\sigma_1) = \begin{pmatrix} -1 & a \\ 0 & q \end{pmatrix} \quad , \quad \mathfrak{F}(\sigma_2) = \begin{pmatrix} q & 0 \\ b & -1 \end{pmatrix} \quad ,$$

where $ab = n = 1 + q - \lambda$. The rational field is a splitting field for the algebra and we have

$$\text{spec}_{\mathbb{Q}}\,(G) = \begin{pmatrix} \text{ind} & \text{st} & \mathfrak{F} \\ 1 & v(q-1)+1 & v-1 \end{pmatrix} \quad .$$

With this information, we can now readily establish a deep application of concept III, the Bruck - Ryser - Chowla - Theorem [7], [10]:

<u>Theorem 4.4</u> <u>Let</u> $G = (\Omega_1, \Omega_2; J)$ <u>be a symmetric design, with parameters</u> $v, k = 1+q, \lambda$. <u>If the number of points is even, then the order</u> $n = 1+q-\lambda$ <u>is a square. If the number of points is odd, then</u>

$$n x^2 + (-1)^{\frac{v-1}{2}} \lambda y^2 = z^2$$

<u>has a nontrivial integral solution.</u>

<u>Proof.</u> Suppose that n is not a square. We determine the discriminant of the eigenspace

$$V_{\mathbb{Q}}(\sigma_i) = \{ f \in V_{\mathbb{Q}} \mid \sigma_i f = \underline{c} f \}$$

by the concept of looking down to the rational field using the field $k = \mathbb{Q}[\sqrt{n}]$ and the eigenspace

$$V_k(\sigma_i) = \{ f \in V_k \mid \sigma_i f = \underline{c} f \} \supseteq V_{\mathbb{Q}}(\sigma_i) \quad .$$

Let $\bar{\ } : x \to \bar{x}$ denote the Galois automorphism of k . The Galois automorphism induces a semilinear map $\Lambda : V_k \to V_k$, which is defined by the rule (with $E \in F$)

$$(\Lambda f)(E) = \overline{f(E)} \quad .$$

By the preceding theorem we know that the standard module allows an orthogonal decomposition

(5) $\quad V_k = V_1 \perp V_2 \perp W \quad ,$

where $V_1 = \{ f \in V_k \mid \sigma_1 f = \sigma_2 f = q f \}$

and $V_2 = \{ f \in V_k \mid \sigma_1 f = \sigma_2 f = -f \}$. Note that $V_1 = \langle \iota \rangle$ with $\iota(F) = 1$.

Because $\Lambda\sigma_i = \sigma_i\Lambda$ and $(\Lambda f, \Lambda g) = \overline{(f,g)}$, we may assume that $\Lambda(W) = W$. Now let $W(\sigma_i) = V_k(\sigma_i) \cap W$. Then

$$V_k(\sigma_1) = V_1 \perp W(\sigma_1) \quad .$$

Consequently we obtain the decomposition

$$(6) \qquad V_{\mathbb{Q}}(\sigma_1) = \langle \iota \rangle \perp W_{\mathbb{Q}}(\sigma_1) \quad ,$$

where

$$W_{\mathbb{Q}}(\sigma_1) = C_{W(\sigma_1)}(\Lambda) = \{ f \in W(\sigma_1) \mid \Lambda f = f \} \quad .$$

We use $d(U)$ to denote the discriminant of a subspace of $V_{\mathbb{Q}}$ and we shall prove

$$(7) \qquad d(V_{\mathbb{Q}}(\sigma_1)) \equiv (1+q)^v \pmod{(\mathbb{Q}^x)^2} \quad .$$

In fact, the equation $(\sigma_1 + 1)(\sigma_1 - q) = 0$ implies that $(\sigma_1 + 1)V_{\mathbb{Q}} = V_{\mathbb{Q}}(\sigma_1)$. Thus, in particular, there exists a basis $\{ \chi_a \mid a \in \Omega_2 \}$ of $V_{\mathbb{Q}}(\sigma_1)$ in which $\chi_a = (\sigma_1 + 1)\chi_{\{A,a\}}$ for an arbitrary $A \, J \, a$. It is clear that $(\chi_a, \chi_b) = (1+q)\delta_{ab}$. Thus (7) holds. Moreover, the preceding argument shows

$$(8) \qquad \dim V_k(\sigma_1) = v \quad .$$

On the other hand, we now claim that $v \equiv 1 \pmod 2$ and

$$d(W_{\mathbb{Q}}(\sigma_1)) \equiv n^{\frac{v-1}{2}} \operatorname{Norm}(z) \pmod{(\mathbb{Q}^{\times})^2} \tag{9}$$

for a suitable $z \in k$.

We begin with the observation (using the relations mentioned in the preceding theorem and the decomposition (5)) that the element

$$\eta = (\lambda - 1)\sigma_1 + \sigma_1\sigma_2\sigma_1 = (\lambda - 1)\sigma_2 + \sigma_2\sigma_1\sigma_2$$

is a self-adjoint unit $(\bar{\eta} = \eta)$, and

$$\begin{aligned} \eta^2_{|W} &= q^2 n\, 1_W \\ \eta\sigma_1 &= \sigma_2\eta \end{aligned} \tag{10}$$

We have therefore

$$\begin{aligned} \eta W(\sigma_2) &= W(\sigma_1) \\ \eta W(\sigma_1) &= W(\sigma_2) \end{aligned}.$$

Now we define a linear map $\alpha: W(\sigma_1) \to W(\sigma_1)$ as follows: Let π denote the orthogonal projection of V_k onto $W(\sigma_2)$, and set

$$\alpha = \eta_{|W(\sigma_2)}\pi .$$

Then by definition

$$(\alpha x, y) = (x, \eta y) \qquad x, y \in W(\sigma_1) .$$

Furthermore, it is easily verified that $\Lambda\pi = \pi\Lambda$, hence $\Lambda\alpha = \alpha\Lambda$. From (10) , we then deduce that α is a self-adjoint map, and

$$\alpha^2 = q^2 n\, 1_{W(\sigma_1)} .$$

Therefore we have shown that the linear map α has exactly the two eigenvalues $\pm q\sqrt{n}$. This yields the important decomposition

$$W(\sigma_1) = W^+ \perp W^- ,$$

where $W^\pm$ is the eigenspace of α corresponding to the eigenvalue $\pm q\sqrt{n}$. Sine $\Lambda\alpha = \alpha\Lambda$, it follows at once

$$\Lambda W^+ = W^-$$
$$\Lambda W^- = W^+ .$$

Comparison with (8) yields

$$\dim W(\sigma_1) = \dim V_k(\sigma_1) - 1 = V - 1 \equiv 0 \quad (\text{mod } 2) .$$

We can prove now the statement (9) by an argument in complete analogy to the proof of theorem 2.3.

Using the fact that $(\iota,\iota) = v(1+q)$ and $v \equiv 1 \pmod 2$, we have by (9) and (7) ,

(11) $$v \equiv n^{\frac{v-1}{2}} \operatorname{Norm}(z) \quad (\text{mod } (\mathbb{Q}^x)^2)$$
for a suitable $z \in k$.

The well-known formula $(v-1)\lambda = q(1+q)$ implies that $v\lambda = (q+1)^2 - n$, hence

$$v \equiv \lambda \operatorname{Norm}(z_o) \quad (\text{mod } (\mathbb{Q}^x)^2)$$
for a suitable $z_o \in k$.

Thus the expression in (11) becomes

$$\lambda \equiv n^{\frac{v-1}{2}} \operatorname{Norm}(z_1) \quad (\text{mod } (\mathbb{Q}^x)^2)$$
for a suitable $z_1 \in k$.

However, this congruence is equivalent to the existence of a nontrivial integral solution of

$$n x^2 + (-1)^{\frac{v-1}{2}} \lambda y^2 = z^2 \quad ,$$

and the proof of the theorem is completed. □

Similar results may be found in Ott [27] .

5. Representations and applications

In this section we shall present a few results on representations. Detailed proofs of these and other results on Hecke algebras may be found in [27].

In the following, G denotes a finite regular geometry of rank $n \geq 2$, with parameters $q_1, q_2, \ldots, q_n \geq 1$. For simplicity, we may assume that the (flag graph of the) geometry G is connected. From the assumption of regularity we deduce the obvious fact that

$$(1) \qquad \sigma_i^2 = q_i \cdot 1 + (q_i - 1)\sigma_i \quad .$$

We observe at once that

$$e_i = \frac{\sigma_i + 1}{q_i + 1}$$

is an idempotent. The corresponding algebra $H_i = e_i H e_i$ is called the <u>adjacency algebra of type</u> i.

As a remark, we turn our attention to an interesting problem concerning graphs. A graph G with vertex-set Ω_1 can be viewed in a natural way as a geometry $G = (\Omega_1, \Omega_2, J)$ of rank $n = 2$. One question immediately arises: Is there a relation between the Hecke algebra and the ordinary adjacency algebra of G? The answer is the following:

<u>Theorem 5.1</u> *Let $G = (\Omega_1, \Omega_2; J)$ be a regular graph of valency $\zeta \geq 2$. Then the adjacency algebra of type 2 is isomorphic*

to the ordinary adjacency algebra of G .

Returning to representations, our first task is to show the existence of two special one-dimensional representations. The simplest is the representation given by

Lemma 5.2 *There is an algebra homomorphism* ind: $H \to k$ *which maps* σ_i *to* q_i .

Proof. We have $\sigma_i \iota = q_i \iota$.

Since G is connected, the following corollary is obvious.

Corollary 5.3 *The multiplicity of* ind *is one*.

For the second one-dimensional representation, we recall that the Steinberg module of G is the factor space of V by the proper subspace $B = B_n \subset V$ of n-boundaries. An easy computation shows that B is σ_i-invariant, the action of σ_i on $H_n = V/B$ being given as multiplication by -1 . Using this, we have the following result:

Lemma 5.4 *There is an algebra homomorphism* st : $H \to k$ *which maps* σ_i *to* -1 .

We easily obtain

(2) $H_n \cong St(G) = \{ f \in V \mid \sigma_i f = -f \}$.

As we have indicated in section 3 , the Euler - Poincaré characteristic can be used in calculating the multiplicity of the representation st in some easy cases. But in general, usually the calculation of multiplicities rests on orthogonality relations. As in section 2 , we introduce the bilinear form (,): $H \times H \to k$ defined by the rule

$$(3) \quad (x,y) = \frac{1}{|F|} \Phi(xy), \qquad x,y \in H.$$

This bilinear form is obviously associative and symmetric. Since the Hecke algebra is semi - simple by theorem 4.3, we deduce

<u>Theorem 5.5</u> <u>The Hecke algebra is a symmetric algebra with respect to the bilinear form given by</u> (3) .

With respect to dual bases $\mathfrak{A} = \{\alpha_1 = 1, \alpha_2, \ldots, \alpha_s\}$ and $\mathfrak{B} = \{\beta_1 = 1, \beta_2, \ldots, \beta_s\}$ of the Hecke algebra H we define a bilinear form on the dual space H^* by the formula

$$(4) \quad [\psi,\chi] = \sum_{j=1}^{s} \psi(\alpha_j)\,\chi(\beta_j), \qquad \psi,\chi \in H^*.$$

We obtain the fundamental orthogonality relations:

<u>Lemma 5.6</u> Let ψ,χ be the characters of two irreducible non - isomorphic H - modules. Then $[\psi,\chi] = 0$.

As a simple consequence of the orthogonality relations, we give the formula for the multiplicities:

(5) $$n_i = \frac{|F|\,\chi_i(1)}{[\chi_i, \chi_i]}, \qquad i = 1, 2, \dots, r.$$

We cannot give any straightforward general method for obtaining dual bases of the Hecke algebra, suitable for character calculations. The Hecke algebra of a geometry of LIE type $(A_n, C_n, D_n, E_6, E_7, E_8, F_4)$ or a generalized m-gon (o—m—o), however, admits a natural pair of dual bases: The structure of a geometry of LIE type or a generalized m-gon can be described by means of a Coxeter diagram over $\{1,2,\dots,n\}$ (Tits [35]) and a finite Coxeter system (W,R), with set of distinguished generators $R = \{r_1, r_2, \dots, r_n\}$. For each reduced word

$$w = r_{i_1} r_{i_2} \dots r_{i_k}$$

in the Weyl group W, let

$$\Lambda(w) = \sigma_{i_1} \sigma_{i_2} \dots \sigma_{i_n} .$$

By well-known properties of Coxeter systems, this definition makes sense. Now we obtain our desired pair of dual bases:

(6) $$\mathfrak{A} = \{ \alpha_w = \Lambda(w) \mid w \in W \}$$
$$\mathfrak{B} = \{ \beta_w = \frac{\overline{\Lambda(w)}}{\operatorname{ind}(\Lambda(w))} \mid w \in W \} .$$

It is of great interest that the construction of the dual bases can be reduced to homotopy-properties of the corresponding Coxeter system. It is possible, abstracting from the geometry, to introduce the general notion of a Coxeter algebra. This concept provides the appropriate method for investigating the universal cover of a finite Tits geometry; Tits [35], Ott - Ronan [23], Ott [24].

As a first illustration of the preceding methods, we mention the theorem of Feit - Higman [14]. The proof of this theorem which we shall sketch now is the celebrated simplified one due to Kilmoyer and Solomon [22]: As we have pointed out, the dimension of the Hecke algebra $H_C(G) = H$ of a finite generalized n - gon G equals the order of the corresponding Weyl group. Thus $\dim_C H = 2n$ and H has defining relations

(7)
$$\sigma_i^2 = q_i + (q_i - 1)\sigma_i, \qquad i = 1,2$$
$$(\sigma_1\sigma_2)^m = (\sigma_2\sigma_1)^m, \qquad \text{if} \quad n = 2m$$
resp.
$$(\sigma_1\sigma_2)^m\sigma_1 = (\sigma_2\sigma_1)^m\sigma_2, \qquad \text{if} \quad n = 2m + 1 .$$

The knowledge of this complete set of relations which the generators σ_1, σ_2 satisfy can be used to construct all irreducible H - modules. Apart from one - dimensional representations, there are only two - dimensional irreducible representations which map

$$\sigma_1 \quad \text{to} \quad \begin{pmatrix} -1 & a \\ 0 & q_1 \end{pmatrix}$$

and

$$\sigma_2 \quad \text{to} \quad \begin{pmatrix} q_2 & 0 \\ b & -1 \end{pmatrix}$$

for suitable $a, b \in C$. As was discussed in section 2 for Moore graphs, the argument that by (5)

$$[\chi_i, \chi_i] = \sum_{w \in D_n} \frac{1}{\text{ind}(\Lambda(w))} \chi(\Lambda(w))\, \chi(\overline{\Lambda(w)})$$

is a rational number implies the main result

<u>Theorem 5.7</u> <u>Let</u> G <u>be a generalized n-gon, with parameters</u> $q_1, q_2, q_1q_2 > 1$. <u>Then</u> $n \in \{2, 3, 4, 6, 8, 12\}$.

We can apply this rather simple machinery to obtain a further interesting result concerning generalized hexagons. If a finite generalized hexagon admits a polarity then the standard module of the hexagon may be viewed as a module for the Hecke algebra of a suitable generalized 12-gon. Again, the application of (5) yields the following result (Ott [25]):

<u>Theorem 5.8</u> <u>If a generalized hexagon of order</u> $q \neq 1$ <u>allows a polarity, then the integer</u> $3q$ <u>is a square</u>.

On the basis of the Kilmoyer - Solomon proof of the Feit - Higman theorem we are able to give a generalization of this result to quasi - n - gons or Moore geometries ([26], compare also with Damerell, Georgiacodis [12], Damerell [13], Fuglister [15]).
Now let $n = 2m + 1 \geq 3$ be an odd integer. A finite geometry G of (point-) diameter m, with parameter q_1 and q_2, $q_1q_2 > 1$, is called a <u>quasi - n - gon</u> or a <u>Moore geometry</u> if each pair of points of the geometry is joined by a unique shortest path of length at most m. We should note that Buekenhout [5] introduces a more general notion of a (g, d^*, d)-gon and this subject becomes an attractive one for study.

Generalized n - gons are examples of quasi - n - gons. By the Feit - Higman theorem, it is sufficient to study <u>proper</u> quasi - n - gons. In this case, we have

$$\dim_C H_C(G) = 2n + 1 \quad .$$

The structure of H cannot be characterized by means of generators and relations, suitable for character calculations. To be effective we must weaken the structure of the Hecke algebra by introducing an algebra A over the field C with generators Γ_1 and Γ_2, and defining relations

$$\Gamma_i^2 = q_i \cdot 1 + (q_i - 1)\Gamma_i \quad , \quad i = 1,2$$

(8) $$(1 + \Gamma_2)(\Gamma_1\Gamma_2)^m \Gamma_1 (1 + \Gamma_2) =$$

$$q_1 q_2 (1 + \Gamma_2)(\Gamma_1\Gamma_2)^{m-1} \Gamma_1 (1 + \Gamma_2)$$

Again, apart from one-dimensional representations, the algebra A admits only two-dimensional irreducible representations which map

$$\Gamma_1 \quad \text{to} \quad \begin{pmatrix} -1 & a \\ 0 & q_1 \end{pmatrix}$$

and

$$\Gamma_2 \quad \text{to} \quad \begin{pmatrix} q_2 & 0 \\ b & -1 \end{pmatrix}$$

where $a\,b = q_1 + q_2 + 2\sqrt{q_1 q_2} \cos\alpha$

and $0 < \alpha < \pi$ with

$$\sqrt{\frac{q_2}{q_1}} \sin(m + 1)\alpha + \sin m\alpha = 0 \quad .$$

The mentioned methods of character calculations are then applied to obtain the major result

Theorem 5.9 Let G be a quasi-n-gon, with parameters q_1 and q_2, $q_1q_2 > 1$. Then $n = 3$ or $n = 5$.

We conclude this section with a remark on coherent configurations. The theory of coherent configurations was developed by Higman [19], [20] to provide the combinatorial foundation of permutation group theory. Therefore it is evident that some geometries have such a rigid structure (Higman [18]). For example, in dealing with generalized polygons or geometries of LIE type, one obtains a homogenous coherent configuration by introducing the relation R_w, $w \in W$, which is defined by means of

$$F R_w G, \text{ if } (\Lambda(w) F, G) \neq 0 , \quad F, G \in F .$$

Furthermore, one can easily verify that the adjacency algebra of this configuration is isomorphic to the Hecke algebra. However, one cannot assert that there is a general connection. For instance, it does not seem to be possible to find a coherent configuration for λ-designs with $\lambda \geqslant 2$.

REFERENCES

[1] BAER, R. Polarities in finite projective planes. Bull. Amer. Math. Soc. 52 (1946) 77 - 93

[2] BALL, R.W. Dualities of finite projective planes. Duke Math. J. 15 (1948) 929 - 940

[3] BANNAI, E. , ITO, T. On finite Moore graphs. J. Fac. Sc. Univ. Tokyo 20 (1973) 191 - 208

[4] BIGGS, N.L. Algebraic Graph Theory. Cambridge Math. Tracts 67, Cambridge Univ. Press (1974)

[5] BUEKENHOUT, F. (g, d*, d) - gons. (to appear)

[6] BUEKENHOUT, F. The basic diagram of a geometry. This volume.

[7] BRUCK, R.H. , RYSER, H.J. The nonexistence of certain finite projective planes. Canad. J. Math. 1 (1949) 88 - 93

[8] BRUCK, R.H. Difference sets in a finite group. Trans. Amer. Math. Soc. 78 (1955) 464 - 481

[9] CAMERON, P.J. Strongly regular graphs. Selected Topics in Graph Theory. Academic Press, London, New York, San Francisco (1978) 337 - 360

[10] CHOWLA, S. , RYSER, H.J. Combinatorial problems. Canad. J. Math. 2 (1950) 93 - 99

[11] DAMERELL, R.M. On Moore graphs. Proc. Cambr. Phil. Soc. 74 (1973) 227 - 236

[12] DAMERELL, R.M. , GEORGIACODIS, M.A. On Moore Geometries I (to appear)

[13] DAMERELL, R.M. On Moore Geometries II (to appear)

[14] FEIT, W. , HIGMAN, G. The nonexistence of certain generalized polygons. J. Alg. 1 (1964) 114 - 131

[15] FUGLISTER, F.J. On finite Moore geometries. J. Comb. Th. 23 (1977) 187 - 197

[16] HALL, M. Cyclic projective planes. Duke Math. J. 14 (1947) 1079 - 1090

[17] HIGMAN, D.G. Partial geometries, generalized quadrangles and strongly regular graphs. Atti del Conv. Geo. Comb. Perugia (1971) 265 - 293

[18] HIGMAN, D.G. Invariant relations, coherent configurations, generalized polygons. Combinatorics, Reidel, Dordrecht (1975) 347 - 363

[19] HIGMAN, D.G. Coherent Configurations. Part I, Ordinary Representation Theory. Geo. Ded. 4 (1975) 1 - 32

[20] HIGMAN, D.G. Coherent Configurations. Part II, Weights. Geo. Ded. 5 (1976) 413 - 424

[21] HOFFMAN, A.J. , NEWMAN, M. , STRAUS, E.G. , TAUSSKY, O. On the number of absolute points of a correlation. Pacif. J. Math. 6 (1956) 83 - 96

[22] KILMOYER, R. , SOLOMON, L. On the theorem of Feit - Higman. J. Comb. Theory Ser. A, 15 (1973) 310 - 322

[23] OTT, U. , RONAN, M.A. On buildings and locally finite Tits geometries. Finite Geometries and Designs. London Math. Soc. Lecture Note Series 49 (1981) 272 - 274

[24] OTT, U. Bericht über Hecke Algebren und Coxeter Algebren endlicher Geometrien. Finite Geometries and Designs. London Math. Soc. Lecture Note Series 49 (1981) 260 - 271

[25] OTT, U. Eine Bemerkung über Polaritäten eines verallgemeinerten Hexagons. Geo. Ded., in press

[26] OTT, U. Quasi - n - gons (to appear)

[27] OTT, U. Hecke Algebren endlicher Geometrien (to appear)

[28] PAYNE, S.E. Symmetric Representations of Nondegenerate Generalized n - Gons. Proc. Amer. Math. Soc. 19 (1968) 1371 - 1326

[29] RONAN, M.A. Coverings of certain finite geometries. Finite Geometrie and Designs. London Math. Soc. Lecture Note Series 49 (1981) 316 - 331

[30] SCHUETZENBERGER, M.P. A non-existence theorem for an infinite family of symmetrical block designs. Ann. Eugenics 14 (1949) 286 - 287

[31] SEIDEL, J.J. Strongly regular graphs. Surveys in Combinatorics. London Math. Soc. Lecture Note Series 38 (1979) 157 - 180

[32] STEINITZ, E. Beiträge zur Analysis situs. Sitzungsberichte der Berl. Math. Ges. (1908) 29 - 49

[33] STEINITZ, E. Polyeder und Raumeinteilungen. Encyklopädie der Math. Wiss. Teubner, Leipzig (1914 - 1931)

[34] THAS, J.A. On Polarities of Symmetric Partial Geometries. Arch. Math. 25 (1974) 394 - 399

[35] TITS, J. Local characterizations of buildings (to appear)

GEOMETRY AND LOOPS

Karl Strambach

Mathematisches Institut
Universität Erlangen-Nürnberg
Bismarckstraße 1 1/2
D-8520 Erlangen

§ 0 Introduction

This article contains some samples from a bigger project which A. Barlotti and I want to realize. Its aim is to show that classical principles of projective geometry and of the foundations of geometry can be applied successfully for the study of loops.

In the second paragraph natural analogues of the classical fundamental theorem of projective geometry are proved for loops and the abelian groups are characterized within the wide class of loops with the help of the group of projectivities.

In the third paragraph it is shown that the transitivity of the collineation group on the points of the 3-net which is associated to a loop Q is equivalent to the fact that every element of Q is a companion of a right and of a left pseudoautomorphism. The stabilizers of the collineation group on the horizontal line 1_h, on the vertical line 1_v and on the point (1,1) is determined and the algebraic consequences for loops with transitive automorphism groups are discussed.

In the last paragraph we present a classification for loops analogous to the Lenz-Barlotti-classification for projective planes and show that this principle which has been propagated by H. Lenz with great success in the foundations of geometry can also be applied for other classes of mathematical structures.

§ 1. Quasigroups, nets and projectivities

Definition (1.1): A k-net (≥ 3) is a structure consisting of a set P of points and a set of lines which is partitioned into k disjoint families $L_i (i = 1,\ldots k)$ for which the following conditions hold:

i) every point is incident with exactly one line of every $L_i (i = 1,\ldots k)$;

ii) two lines of different families have exactly one point in common;

iii) there exist 3 lines belonging to 3 different L_i and which are not incident with the same point.

Lines of the same [different] families are said to have the same [different] directions.

It is well known that to every quasigroup Q (see e.g. [8], p. 16) we can associate a 3-net (see e.g. [8], p. 251) such that the three families of parallel lines consist of the following sets of points:

$g_h = \{(x, g) |$ g constant, $x \in Q \}$, horizontal lines;

$g_v = \{(g, x) |$ g constant, $x \in Q \}$, vertical lines;

$g_t = \{(x, y) |$ $x \cdot y = g$; $x, y \in Q$, g constant$\}$, transversal lines.

We shall denote by $\mathfrak{H}$, $\mathfrak{V}$ and $\mathfrak{T}$ the families of horizontal, vertical and transversal lines respectively.

Conversely every 3-net leads to a class of isotopic quasigroups (see e.g. [6], p. 20).

Let N be a k-net, L a line in it, and $\mathfrak{X}$ one of the k families of parallel lines such that $L \notin \mathfrak{X}$. A perspectivity $\alpha = [L, \mathfrak{X}]$ assigns to a point $x \in L$ the line X of $\mathfrak{X}$ through x . The perspectivity $\alpha^{-1} = [\mathfrak{X}, L]$ assigns to $X \in \mathfrak{X}$ the point $x = X \cap L$.

A <u>projectivity</u> γ of a line onto a line is given by a set of consecutive perspectivities α_i, or in other words γ is the product of these α_i:

$$\gamma = \prod_{i=1}^{n} \alpha_i .$$

The projectivities of a line L onto itself in a k-net N form a group Π_L with respect to the composition of mappings. If H is any other line in N and β any projectivity from L onto H, then we have $\Pi_H = \beta^{-1}\Pi_L\beta$. Therefore all groups of projectivities of a line onto itself in a k-net are isomorphic as permutation groups, and we can speak of the <u>group of projectivities</u> of N .

If Q is a quasigroup then we define the group Π of projectivities of Q as the group of projectivities in a 3-net N which naturally arises from Q . Clearly all the members of the isotopy class of Q - as quasigroups corresponding to N - have Π as group of projectivities. The same holds even for all quasigroups isostrophic to Q ([3], p. 13).

Therefore Π can be seen as a group of projectivities of an isotrophy class of quasigroups. In any isotopy class of quasigroups there are loops: if (Q, $\cdot$) is a quasigroup and a, b are fixed elements of Q , then it is well known that (Q,$*$), with $(x\cdot a)*(b\cdot y) = x\cdot y$, is a loop, with the identity $b\cdot a$, and is isotopic to (Q,$\cdot$). Therefore it is enough to study the group of projectivities for loops.

Every projectivity α of a line L onto itself in a k-net N can be represented as

$$\overline{\alpha} = \prod_{i=1}^{n} [L_{i-1}, \mathfrak{X}_i][\mathfrak{X}_i, L_i]$$

with $L_o = L_n = L$.

We say that the representation $\overline{\alpha}$ is <u>irreducible</u> (of length n) if $L_i \neq L_{i+1}$ and $\mathfrak{X}_i \neq \mathfrak{X}_{i+1}$. To a representation $\overline{\alpha}$ and to the set of points $c = \{ a_i^o \mid i = s,\dots, m;\ a_i^o \in L \}$ we associate the configuration $\Omega(\overline{\alpha}, c)$ consisting of all lines L_i (the "generators" of Ω), of the points

$$a_i^{(k)} = (a_i^o)\overline{\alpha}^k = (a_i^o) \prod_{i=1}^{k} [L_{i-1}, \mathfrak{X}_i] [\mathfrak{X}_i, L_i]$$

and of the lines (the non-trivial "projection lines") joining the different pairs of points $a_i^{(k-1)}$ and $a_i^{(k)}$.

The general problem of determining the group Π of projectivities of a given loop Q seems to be difficult. If however the loop Q satisfies some additional algebraic properties then we can determine the group Π of projectivities explicitly. Now we will compute Π for loops Q having the inverse property. We remember that a loop Q has the <u>inverse property</u> if and only if for every x there exist a and b in Q such that

$$a(xy) = y \quad \text{and} \quad (yx)b = y \quad \text{for all} \quad y \in Q$$

[6], p. 111.

If G is a loop with the inverse property, then we will call $P(G)$ the group which is generated by the mappings

$$\{x \to (ax)b \quad \text{and} \quad x \to a(xb) \; ; \; G \to G\}.$$

Let L respectively R be the set of all left translations $x \to ax$, respectively of all right translations $x \to xa$. Let $L(G)$ respectively $R(G)$ be the group generated by L respectively R. If G is a group then $P(G)$ is a product of $L(G)$ and $R(G)$; moreover $L(G)$ and $R(G)$ are then normal subgroups of G.

In any case $L \cap R$ consists of translations $x \to ax$ such that a is contained in the centre of G; that is $ax = xa$ for all x. This assertion follows from the fact that if a left multiplication $x \to ax$ belongs to R, then there exists an element b such that for every x the equation $ax = xb$ holds; therefore $(ax)b^{-1} = x$, since a loop with the inverse property has a unique inverse element (see [6], p. 111). For $x = 1$ it follows $ab^{-1} = 1$ and so $b=a$. Then we have $ax = xa$ for all $x \in G$.

If in particular G is a group then, moreover $R(G) \cap L(G)$ consists of all translations with elements out of the centre of G. Moreover in $P(G)$ every element of $L(G)$ commutes with every element of $R(G)$; since $(g_1x)g_2 = g_1(xg_2)$ holds for all $x \in G$ we have for the left translation λ_{g_1} and the right translation ρ_{g_2} the equation

$$\lambda_{g_1} \rho_{g_2} = \rho_{g_2} \lambda_{g_1} .$$

Theorem (1.1). The group Π *of projectivities of a loop* G *with the inverse property is isomorphic as permutation group to the group generated by the group* P(G) *and the mapping* $\tau = (x \to x^{-1} : G \to G)$. *The group* P(G) *is a normal subgroup of* Π .

The mapping τ *operates on* $L \cup R \subseteq P(G)$ *in the following way. For* $\rho_a = (x \to xa)$ and $\lambda_a = (x \to ax)$ *holds* $\tau\rho_a\tau = \lambda_{a^{-1}}$ and $\tau\lambda_a\tau = \rho_{a^{-1}}$.

If $\tau \notin P(G)$ *- and this for instance is the case when* G *is a group - then* Π *is the semidirect product of* P(G) *and* $<\tau>$.

Proof. If N(G) is the 3-net associated to G then we can describe the action of the different types of perspectivities within N(G) as follows.

Perspectivity	Preimage	Image
$[g_h, \mathfrak{V}]$	the point (x, g)	the line x_v
$[g_h, \mathfrak{V}]^{-1} = [\mathfrak{V}, g_h]$	the line x_v	the point (x, g)
$[g_h, \mathfrak{T}]$	the point (x, g)	the line $(xg)_t$
$[g_h, \mathfrak{T}]^{-1} = [\mathfrak{T}, g_h]$	the lines x_t	the point (xg^{-1}, g)
$[g_v, \mathfrak{H}]$	the point (g, x)	the line x_h
$[g_v, \mathfrak{H}]^{-1} = [\mathfrak{H}, g_v]$	the line x_h	the point (g, x)
$[g_v, \mathfrak{T}]$	the point (g, x)	the line $(gx)_t$
$[g_v, \mathfrak{T}]^{-1} = [\mathfrak{T}, g_v]$	the line x_t	the point $(g, g^{-1}x)$
$[g_t, \mathfrak{H}]$	the point (x, y) (with $xy = g$)	the line y_h
$[g_t, \mathfrak{H}]^{-1} = [\mathfrak{H}, g_t]$	the line x_h	the point (gx^{-1}, x)
$[g_t, \mathfrak{V}]$	the point (x, y) (with $xy = g$)	the line x_v
$[g_t, \mathfrak{V}]^{-1} = [\mathfrak{V}, g_t]$	the line x_v	the point $(x, x^{-1}g)$

We assume now that α is a projectivity of a line g_h onto itself. Then α can be decomposed in projectivities γ_i of smallest length such that the preimage line and the image line of γ_i are always in the set $\mathfrak{H} \cup \mathfrak{V}$. We discuss now the possibilities for γ_i and describe the action on the points explicitly .

$$\left.\begin{array}{ll}
\text{If } \gamma_i = [g_h, \mathfrak{v}][\mathfrak{v}, g_h'] & \text{then } (x, g)\gamma_i \to (x, g');\\
\text{if } \gamma_i = [g_h, \mathfrak{z}][\mathfrak{z}, g_h'] & \text{then } (x, g)\gamma_i \to [(xg)g'^{-1}, g'];\\
\text{if } \gamma_i = [g_v, \mathfrak{h}][\mathfrak{h}, g_v'] & \text{then } (g, x)\gamma_i \to (g', x);\\
\text{if } \gamma_i = [g_v, \mathfrak{z}][\mathfrak{z}, g_v'] & \text{then } (g, x)\gamma_i \to [g', g'^{-1}(g\,x)];\\
\text{if } \gamma_i = [g_h, \mathfrak{z}][\mathfrak{z}, g_v'] & \text{then } (x, g)\gamma_i \to [g', g'^{-1}(xg)];\\
\text{if } \gamma_i = [g_v, \mathfrak{z}][\mathfrak{z}, g_h'] & \text{then } (g, x)\gamma_i \to [(gx)g'^{-1}, g'];\\
\text{if } \gamma_i = [g_h, \mathfrak{v}][\mathfrak{v}, g_t'][g_t', \mathfrak{h}][\mathfrak{h}, g_v''] & \\
 & \text{then } (x, g)\gamma_i \to (g'', x^{-1}g');\\
\text{if } \gamma_i = [g_v, \mathfrak{h}][\mathfrak{h}, g_t'][g_t', \mathfrak{v}][\mathfrak{v}, g_h''] & \\
 & \text{then } (g, x)\gamma_i \to (g'x^{-1}, g'').
\end{array}\right\} \quad (1)$$

If one piece of a projectivity is of the form:

$$[g_h, \mathfrak{v}][\mathfrak{v}, g_t^{(1)}][g_t^{(1)}, \mathfrak{h}][\mathfrak{h}, g_t^{(2)}] \ldots\ldots$$

or $[g_v, \mathfrak{h}][\mathfrak{h}, g_t^{(1)}][g_t^{(1)}, \mathfrak{v}][\mathfrak{v}, g_t^{(2)}] \ldots\ldots$

then this can be written in (reducible) form as

$$[g_h, \mathfrak{v}][\mathfrak{v}, g_t^{(1)}][g_t^{(1)}, \mathfrak{h}][\mathfrak{h}, g_v^{*}][g_v^{*}, \mathfrak{h}][\mathfrak{h}, g_t^{(2)}] \ldots \quad (2)$$

or respectively

$$[g_v, \mathfrak{h}][\mathfrak{h}, g_t^{(1)}][g_t^{(1)}, \mathfrak{v}][\mathfrak{v}, g_h^{*}][g_h^{*}, \mathfrak{v}][\mathfrak{v}, g_t^{(2)}] \ldots \quad (3)$$

Since (2) and (3) can be expressed as products of the last two projectivities given in (1) it is clear that the projectivity α can be decomposed in projectivities γ_i such that every γ_i occurs in (1). Every γ_i acts only on the variable coordinate x. The image of x arises under each γ_i by a suitable composition of the following mappings:

$$x \to a\,x, \qquad x \to x\,b, \qquad x \to x^{-1}$$

Therefore α has the same property and the theorem follows since for every two elemente c, d we have $(cd)^{-1} = d^{-1}c^{-1}$ [6], p. 111, (1.8)). □

Let N be a 3-net which is embedded in an affine plane A. It should be noticed that not every collineation of A leaving N invariant induces a projectivity of N. In the classical cases also if A is desarguesian and N is the additive 3-net (cf. [21], p. 61) besides the translations of A the only collineations which induce projectivities are the reflections on a point.

At the end of this section we give examples of loops G with the inverse property such that the map $\tau = (x \to x^{-1}; G \to G)$ is contained in $P(G)$.

Let Q_m be the free loop over a set of generators with the cardinality m. Let us denote by x_l^{-1}, respectively x_r^{-1} the elements of Q_m defined by $x_l^{-1}x = 1 = xx_r^{-1}$. Let N be the normal subloop belonging to the relations

$$y_l^{-1}[(yx)x_r^{-1}] = 1 , \quad [x_l^{-1}(xy)]y_r^{-1} = 1 .$$

The factor loop $\Psi_m = Q_m/N$ is the free loop with the inverse property over a set of generators with the cardinality m. In Ψ_m holds $x_l^{-1} = x_r^{-1} = x^{-1}$ ([6], p. 111). Let now M be the normal subloop of Ψ_m belonging to the relation $[y^{-1}(xy)]x=1$. It is clear that in the factor loop $\Phi_m = \Psi_m/M$ the map τ is contained in $P(\Phi_m)$. Since the loop Φ_m is not power-associative the stabilizer of $\Pi(\Phi_m)$ on the point $(1,1)$ of the line 1_h contains - besides τ - many other elements different from the identity, for instance the maps $\tau_a = [x \to (a^{-1}x)a : \Phi_m \to \Phi_m]$ which are all different from τ.

§ 2. The Staudt's Theorems for loops

In this section we give geometric characterizations for the abelian groups in the wide class of loops.

Contrary to the case of projective planes there is no chance of characterizing the whole class of abelian groups by the condition that the stabilizer $\Pi_{x_1,\ldots,x_n}$ of the group of projectivities Π fixing every element of an arbitrary n-tuple consists only of the identity. In fact let G be an abelian group which is not an elementary 2-group and which has s involutions. Then $x \to x^{-1}$ is a projectivity $\neq 1$ having $s+1$ fixed points. However for the class of abelian groups without involutions we have a direct analogue of the classical Staudt theorem for planes.

Theorem (2.1). A loop G is an abelian group without involutions if and only if the pointwise stabilizer of Π on every two distinct points consists only of the identity.

Before we give the proof of theorem (2.1) we notice the following

Proposition (2.2). If an a 3-net N all those Thomsen configurations close for which the three diagonals do not intersect at the same point then the hexagonal condition holds (i.e. the Thomsen condition holds without restrictions).

Proof.

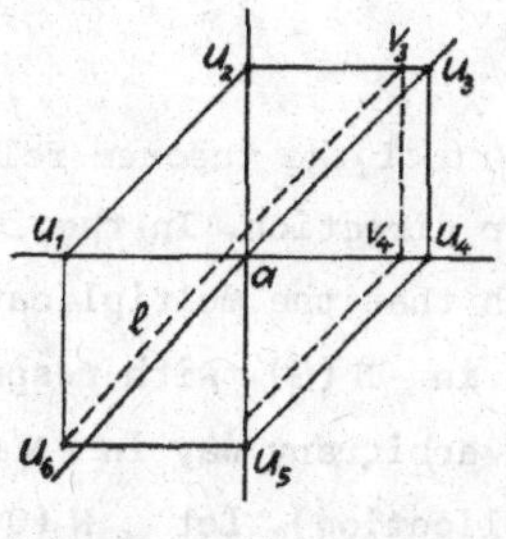

If the hexagonal condition with respect to the point a does not hold then the point u_6 in figure 1 does not belong to the line au_3. Then the line ℓ through u_6 which belongs to the same family $\mathfrak{Z}$ as au_3 intersects the two lines through a of the other two families in two different points, and meets the line u_2u_3 in a point $v_3(\neq u_3, u_2)$. Let v_4 be the intersection of u_1u_4 with the line through v_3 belonging to the same family as au_2. The points u_6, u_1, u_2, v_3, v_4 and the lines u_1v_4, u_2a

and ℓ lead to a Thomsen configuration which satisfies our assumption but which does not close. □

Proof of theorem (2.1). Let us consider the assumptions of the Thomsen condition. Let V_1, V_2, V_3 be three different vertical lines and H_1, H_2, H_3 be three different horizontal lines and assume that the points $V_1 \cap H_1$ and $V_3 \cap H_3$ are on a transversal line T_1 and that the points $H_1 \cap V_2$ and $H_2 \cap V_3$ are on a transversal line T_2. Under these hypotheses we want to prove that the points $H_2 \cap V_1$ and $H_3 \cap V_2$ are also on a transversal T_3. We can (assume because of proposition (2.2)) that the two points $a = H_2 \cap V_2$ and $b = H_2 \cap T_1$ are different. Consider then the following projectivity:

$$\delta = [H_2, \mathfrak{V}][\mathfrak{V}, T_1][T_1, \mathfrak{h}][\mathfrak{h}, V_2][V_2, \mathfrak{z}][\mathfrak{z}, H_2]$$

The projectivity δ^2 fixes the points a and b; therefore $\delta^2 = 1$. Let be $x = H_2 \cap V_1$ then $x^{\delta^2} = x$ and therefore the transversal line T_3 through x carries the point $V_2 \cap H_3$. Hence the Thomsen condition holds and Q is an abelian group.

Viceversa if Q is an abelian group then the stabilizer of Π on one point has order at most two (cf. § 1) and the assertion follows. □

In order to characterize the whole class of abelian groups by the group of projectivities we take a condition on the order of a stabilizer of Π on a point.

Theorem (2.3). A loop G is an abelian group if and only if the stabilizer Π_a of a point a in the group Π of projectivities of G has order at most two.

Proof. If G is an abelian group then one part of the theorem follows from theorem (1.1). Let us consider the other direction. In the isotopy class of G there exists a loop Q such that the multiplication of Q is given by the natural multiplication in $N(G)$ with respect to the point $0 = (1,1)$ (which is chosen in an arbitrary way in $N(G)$ but which gives the 1 of the natural multiplication). Let $N(Q)$ be the not corresponding to Q, and let $1_v, 1_h, 1_t$ be the vertical, horizontal and transversal lines passing through the point (1,1). The hexagonal condition for the point (1,1) is equivalent with the fact that every element in Q has exactly one inverse ([21], p. 54). Let $\Pi(Q) = \Pi$ be the group of projectivities of Q and let $\Pi_{(1,1)}$ be the stabilizer of Π on (1,1). The maps $(x, 1) \to (x_r^{-1}, 1)$ and $(x, 1) \to (x_l^{-1}, 1)$ (where x_r^{-1} and x_l^{-1} are respectively the right and the left inverse of x) are given by the projectivities

$$[1_h, \mathfrak{W}]\,[\mathfrak{W}, 1_t][1_t, \mathfrak{H}][\mathfrak{H}, 1_v][1_v, \mathfrak{F}][\mathfrak{F}, 1_h]$$

and respectively

$$[1_h, \mathfrak{F}]\,[\mathfrak{F}, 1_v][1_v, \mathfrak{H}][\mathfrak{H}, 1_t][1_t, \mathfrak{W}][\mathfrak{W}, 1_h]\,.$$

Since $|\Pi_{(1,1)}| \leq 2$ holds, it follows that in N the hexagonal condition holds for the point $0 = (1,1)$.

Since $|\Pi_x| \leq 2$ for every point $x \in N(G)$ and every point x can be chosen as origin for a loop belonging to the isotopy class of G, the hexagonality condition holds in general and the isotopy class contains power associative loops only ([1], thm. 3.5, p. 406).

If $|\Pi_x| = 1$ for one point x then Π is sharply transitive, i.e. $|\Pi_x| = 1$ for every x and in the net the condition of "parallel diagonals" holds and ([21], p. 60) the loop Q is a group such that every element different from 1 is an involution. Therefore Q is an elementary abelian 2-group and the theorem holds.

We assume now that for every x is $|\Pi_x| = 2$, and that Q is a power associative loop. Let G_1, G_2, G_3 be three lines belonging to the same class $\mathfrak{X}$, let a be a point on G_1 and $\mathfrak{Y}$, $\mathfrak{Z}$ the two remaining classes of lines. Let us consider the projectivity:

$$\varepsilon = [G_1, \mathfrak{Y}][\mathfrak{Y}, G_2][G_2, \mathfrak{Z}][\mathfrak{Z}, G_3][G_3, \mathfrak{Y}]\,;$$

the two lines $(a)\varepsilon$ and $(a)[G_1, \mathfrak{Z}]$ have exactly one point s in common. Let G_4 be the line of $\mathfrak{X}$ through s.

The projectivity

$$\delta = \varepsilon[\mathfrak{X}, G_4][G_4, \mathfrak{Z}][\mathfrak{Z}, G_1]$$

is contained in the stabilizer Π_a of the point a.

Let us first choose $a = (1,1)$, $G_i \in \mathfrak{H}$,

$G_1 = 1_h$, $G_3 = g''_h$, $G_2 = G_4 = g'_h$ and $G_1 \neq G_2 \neq G_3 \neq G_4 \neq G_1$

and $\mathfrak{X} = \mathfrak{V}$, $\mathfrak{Z} = \mathfrak{Y}$. Then $\delta \in \Pi_{(1,1)}$ is equivalent to $g' = g'^{-1}g''$.
Since Q is power associative it follows $g'' = g'^2$. For any point $(x, 1) \in G_1$ we have then

$$(x, 1)^{\delta} = (\omega, 1)$$

with $xg' = \eta g'^2$ and $\eta g' = \omega$. Since $|\Pi_{(1,1)}| = 2$ then for every x is either $\omega = x$ or $\omega = x^{-1}$. If it is $\omega = x$ for every x then in Q holds one Bol condition. Assume that there exists some $x \in Q$ with $x^{\delta} = x^{-1} \neq x$. Now if $g'^{-1} \neq g'$ we can choose $x = g'^{-1}$ and then it would follow $\eta = g'^{-2}$ and $g' = \omega = g'^{-1}$, therefore a contradiction. If g' is an involution then we have $xg' = \eta$ and $\eta g' = x^{-1}$ and get for every involution g' and every x the following rule of computation

$$(xg')\,g' = x^{-1} . \qquad (1)$$

Let us choose now on the other hand $a = (1,1)$, $G_i \in \mathfrak{V}$,
$G_1 = 1_v$, $G_3 = g''_v$, $G_2 = G_4 = g'_v$ and $G_1 \neq G_2 \neq G_3 \neq G_1$ and

$\mathfrak{X} = \mathfrak{H}$, $\mathfrak{Z} = \mathfrak{Y}$, then $\delta \in \Pi_{(1,1)}$ is equivalent with $g'' = g'^2$.

For any point $(1, x) \in G_1$ we have then $(1, x)^{\delta} = (1, \omega)$ with $g'x = g'^2\eta$ and $g'\eta = \omega$. An analogous computation as before shows us that we have either a further Bol condition or that $\omega = x^{-1}$ holds for every x and that there exists some $x \in Q$ such that $x^{\delta} \neq x^{-1} = x$. In the second case we obtain for every involution g' and every x the following relation

$$g'(g'x) = x^{-1} . \qquad (2)$$

We choose now $a = (1,1)$, $G_i \in \mathfrak{h}$, $G_1 = 1_h$, $G_2 = g'_h$, $G_3 = g''_h$, $G_2 = g'''_h$ in such a way that all the G_i are different and it holds $\mathfrak{h} = \mathfrak{v}$, $\mathfrak{z} = \mathfrak{f}$. In order to have $\delta \in \Pi_{(1,1)}$ we must have $g' = \zeta g''$, $\zeta g''' = 1$ and therefore $g' = g'''^{-1} g''$.

If $(x, 1) \in G_1$ is any point, then we have $(x, 1)^\delta = (\omega, 1)$ with $x(g'''^{-1} g'') = \eta g''$ and $\eta g''' = \omega$. As before, because of $(x \to x^{-1}) \in \Pi_{(1,1)}$, it follows that $\omega = x^{\pm 1}$. Moreover if we choose g''' involutory we obtain that for $x = g'''$ holds $\omega = g'''$ for $x = g'''$.

Then we have

$$g'''(g''' \; g'') = g'' \tag{3}$$

for every involutory element g''' and every element g'' . But (3) contradicts (2).

Finally we choose $a = (1,1)$, $G_i \in \mathfrak{v}$, $G_1 = 1_v$, $G_2 = g'_v$, $G_3 = g''_v$, $G_4 = g'''_v$ in such a way that all the G_i are different and it holds $\mathfrak{h} = \mathfrak{h}$ and $\mathfrak{z} = \mathfrak{f}$. In order that δ fixes the point (1,1) it is necessary that $g' = g'' \, g'''^{-1}$.

For $(1,g''')$, where g''' is any involution, we obtain by a computation analogous to the above the relation

$$(g'' \; g''')g''' = g'' \tag{4}$$

for every element g'' . But (4) contradicts (1).

Therefore in the loop Q two Bol conditions are fulfilled and so all three Bol conditions. Then the loop Q is a Moufang loop ([21], p. 57-58). Now it follows (see for instance [1], p. 416) that every loop of the isotopy class of Q is a Moufang loop. As a Moufang loop Q is di-associative (see [6], p. 117).

We choose now $a = (1,1)$, $G \in \mathfrak{h}$, $G_1 = 1_h$, $G_2 = g'_h$, $G_3 = g''_h$, $G_4 = g'''_h$ in such a way that all the G_i are different and $\mathfrak{h} = \mathfrak{v}$, $\mathfrak{z} = \mathfrak{f}$. In order that δ fixes the point (1,1) we must have $g' = g'''^{-1} g''$. If in Q the Reidemeister condition is not fulfilled then three follows $(x,1)^\delta = (x^{-1}, 1)$ for every x and there exist x with $x \neq x^{-1}$.

Now we have $x(g'''^{-1}g'') = \eta g''$ and $\eta g''' = x^{-1}$.

With $x = g''$ it follows from these that $g''(g'''^{-1}g'') = \eta g''$ and then $[(g''g'''^{-1})g'']g''^{-1} = (\eta g'')g''^{-1}$ and therefore $\eta = g''g'''^{-1}$ and then $(g''g'''^{-1})g''' = g''^{-1}$ and so $g'' = g''^{-1}$. This gives us a contradiction since we have assumed that there is an element in Q different from its inverse.

Hence in Q the Reidemeister condition holds and Q is a group. From the theorem (1.1) we know that the stabilizer $\Pi_{(1,1)}$ of the group of projectivities consists of the mappings $x \to a^{-1}xa$ for every $a \in Q$ and of the mapping $x \to x^{-1}$ which in our case is different from the identity. If G would not be abelian then the only inner automorphism different from the identity should be the inversion $x \to x^{-1}$, since $\Pi_{(1,1)}$ has the order two; but a group for which $x \to x^{-1}$ is an automorphism is abelian. □

A further very easy characterization of the abelian groups is given by the following

<u>Corollary (2.4).</u> A loop Q is an abelian group if and only if the group Π of projectivities of Q contains a sharply point transitive subgroup of index at most two.

<u>Proof.</u> The stabilizer of a point has order at most two. Hence the result follows from theorem (2.3).

At the end of this section we want to exhibit a theorem which is analogous to the theorems which characterize the pappian planes as planes in which every projectivity can be represented by a chain of length less or equal 4 (cf. [21], p. 139).

<u>*Theorem (2.5)*</u> *A loop* Q *is an abelian group if and only if the group* Π *of projectivities contains a subgroup* N *of index* ≤ 2 *in which every projectivity can be represented by a chain with the length* 0 *or* 4.

<u>Proof.</u> If Q is an abelian group then the group of projectivities

is known (see § 1). For N we can take the group of all maps $x \to ax$.

Let Q be a loop satisfying our conditions. The map $\alpha_\ell : x \to x_\ell^{-1}$ where $1 = x_\ell^{-1}x$ is a projectivity such that (if $\alpha_\ell \neq 1$) its shortest representation has length six. If α_ℓ is the identity then every element of $Q \setminus \{1\}$ is an involution and the net $N(Q)$ satisfies the condition of the parallel diagonals. From [21] p. 60 follows that Q is an elementary abelian 2-group. Hence we may assume that $\alpha \neq 1$ holds. Every projectivity of the line l_h of lenght 4 is either

$$\rho_a : x \to xa \quad , \quad a \neq 1$$

or ρ_a^{-1}. Therefore N operates fixed point free on l_h, i.e. only the identity of N has fixed points. If N would not operate transitively on l_h then N would have (since Π is transitive) two distinct domains D_1 and D_2 of transitivity. Assume for instance $(1,1) \in D_1$. One has $\Pi = N \langle \alpha_\ell \rangle$. Now $(1,1)^{\alpha_\ell} = (1,1)$ and hence $D_1^{\alpha_\ell} = D_1 = D_1^{\Pi}$ which is a contradiction to the transitivity of Π. Therefore N is a sharply transitive normal subgroup of Π. Now corollary (2.4) gives the result. □

§ 3. The collineation group of a loop.

The collineation group Σ of a quasigroup Q is the (full) collineation group of the 3-net $N(Q)$ belonging to Q (*). Therefore Σ is the same for every quasigroup out of the same isotopy class, and we can assume that Q is a loop. The group Σ has a normal subgroup Γ of index ≤ 6 in Σ which maps into itself every class of parallel lines; this subgroup will sometimes be called the group of collineations of $N(Q)$ which preserves the directions. If the group Γ contains a subgroup Ψ which leaves every line out of one given class invariant and operates transitively on the line as a point set, then Q is a group and vice-versa ([4] p. 189). For non-associative loops then Σ cannot contain such a transitive "glide group" and the determination of Σ is difficult (cf. [3], chap. V).

On the other hand the determination of the stabilizers Γ_{1_v}, Γ_{1_h} and $\Gamma_{(1,1)}$ on the lines 1_v, 1_h and the point $(1,1)$ is easy.

A permutation α of a loop Q is called a <u>right</u> respectively <u>left pseudoautomorphism</u> if there exists at least one element c of Q, called a <u>companion</u> of α such that for every x, y

$$(x^{\alpha})(y^{\alpha}c) = (xy)^{\alpha}c \quad \text{respectively}$$

$$(cx^{\alpha})(y^{\alpha}) = c(xy)^{\alpha} \quad \text{holds.}$$

If α is such a one-sided pseudo-automorphism then we have $1^{\alpha} = 1$. If it is clear which class of one-sided pseudo-automorphism is considered or if it does not matter whether a pseudo-automorphism is one-sided or two-sided, we sometimes only use the term pseudo-automorphism and its companions . If Q is a loop with the inverse property then every one-sided pseudo-automorphism is two-sided ([6], p. 113); the same holds naturally for commutative loops.

(*) A collineation is defined to be a permutation of this points of $N(Q)$ such that a line is always mapped on a line.

We have the following

Theorem (3.1). *Let* Q *be a loop and* N(Q) *the net belonging to* Q. *Let us denote by* Γ *the group of all collineations of* N(Q) *which leave the set of horizontal lines and the set of vertical lines invariant. Then the stabilizer* Γ_{1_v} *on the line* 1_v *consists exactly of the set of mappings* $(x, y) \to (x^{\alpha_v}, y^{\alpha_h})$ *where* α_v *is a right pseudo-automorphism of* Q *and* $\alpha_h = \alpha_v \tau_c$, *denoting by* τ_c *the right translation by a companion* c *of* α_v. *Likewise the stabilizer* Γ_{1_h} *consists of the set of mappings* $(x, y) \to (x^{\alpha_v}, y^{\alpha_h})$ *where* α_h *is a left pseudo-automorphism of* Q *and* $\alpha_v = \alpha_h \lambda_c$, *denoting by* λ_c *the left translation by a companion* c *of* α_h.

The stabilizer Γ_{1_v} *(respectively* Γ_{1_h}*) is exactly then transitive on* 1_v *(resp.* 1_h*) if every element* $c \neq 1$ *is a companion of a right (respectively left) pseudo-automorphism of* Q.

If Q *has the inverse property then the stabilizer* Γ_{1_v} *(respectively* Γ_{1_h}*) operates exactly then transitively on* 1_v *(respectively* 1_h*) if* Q *is a Moufang loop.*

If Q *is a group then* Γ *consists of the maps* $(x, y) \to (c\,x^{\alpha}, y^{\alpha} d)$ *where* $c, d \in Q$ *and* α *is an automorphism of* Q.

Proof. Let α be a collineation out of Γ_{1_v} and $(x, y)^{\alpha} = (x^{\alpha_v}, y^{\alpha_h})$. The point (x, y) lies on the line $(xy)_t$; this also contains the point $(1, xy)$. Since α leaves invariant also the set of the transversal lines it follows

$$x^{\alpha_v} y^{\alpha_h} = (1^{\alpha_v})(xy)^{\alpha_h} = (xy)^{\alpha_h}$$

since $\alpha \in \Gamma_{1_v}$. With $x = g$ and $y = 1$ we have $g^{\alpha_v}(1^{\alpha_h}) = g^{\alpha_h}$.

Therefore if we put $1^{\alpha_h} = c$ we have

$$(x^{\alpha_v})(y^{\alpha_v} c) = (xy)^{\alpha_v} c$$

and α_v is a right pseudo-automorphism with companion c.

Conversely let α_v be a right pseudo-automorphism of Q with companion c. We shall prove that the mapping

$$(x,y) \to (x^{\alpha_v}, y^{\alpha_v} c) \qquad (*)$$

is a collineation of $N(Q)$. It is clear that the image of a vertical (respectively horizontal) line is a vertical (respectively horizontal) line. Let now (x, y) be such that $xy = d$ where d is a fixed element. Then $x^{\alpha_v}(y^{\alpha_v} c) = (xy)^{\alpha_v} c = d^{\alpha_v} c$ and so the image of a transversal line is a transversal line and $(*)$ is a collineation belonging to Γ_{1_v} since α_v is a right pseudo-automorphism.

The second statement of the theorem is a trivial consequence of the first since Γ_{1_v} is exactly then transitive to the points of 1_v if $(1,1)$ can be mapped by Γ_{1_v} on every point $(1, x)$ with $x \in Q$.

If Q has the inverse property then the group Γ_{1_v} is transitive on 1_v exactly then when every element is a companion of a pseudo-automorphism. Then the loop Q is isomorphic to the isotopic loop defined by $x * (b^{-1}y) = x \cdot y$ since there is a collineation moving $(1,1)$ to any point of the line 1_v (cf. [21], p. 50). From [6], p. 115, thm. 2.3, follows now that b is a Moufang element. Then every element is a Moufang element and Q is then a Moufang loop ([6], p. 113, lemma 2.2). □

It is interesting to notice that the pseudo-automorphisms and their companions have a deep geometrical meaning and appear in such a natural way in the study of the collineation group of a loop as the above theorem shows. Also the automorphism group $A(Q)$ of a loop Q has a natural geometric interpretation. It is clear that we have a natural injection Φ from $A(Q)$ into the stabilizer $\Gamma_{(1,1)}$ of the direction preserving group Γ, namely

$$\varphi : \alpha \to (\hat{\alpha} = [(x,y) \to (x^{\alpha}, y^{\alpha})]).$$

The following theorem shows that φ is even a bijection.

Theorem (3.2) . *In every loop* Q *the injection* φ *gives an isomorphism between the automorphisms group* A(Q) *and the stabilizer* $\Gamma_{(1,1)}$ *of the direction preserving collineation group* Γ *on the point* (1,1).

Proof. One has $\Gamma_{(1,1)} \subseteq \Gamma_{1_v}$

and

$$\Gamma_{1_v} = \{(x,y) \to (x^{\alpha}, y^{\alpha} c)\}$$

where α is a right pseudo-automorphism of Q and c is a companion of α . If $\lambda \in \Gamma_{(1,1)}$ then we have

$$(1,1)^{\lambda} = (1^{\alpha}, 1^{\alpha} c) = (1, c) = (1,1)$$

i.e., c = 1 and α is an automorphism. □

The full stabilizer of the collineation group Σ on the point (1,1) can be computed for more special classes of loops, namely for loops having the inverse property or - expressing the same property geometrically - if in the corresponding net of Q both Bol condition hold for the point (1,1).

Theorem (3.3). *Let* Q *be a loop having the inverse property. Then the stabilizer* $\Sigma_{(1,1)}$ *of the full collineation group* Σ *of* Q *is the direct product of* $\Gamma_{(1,1)} \cong$ Aut Q *with a group* Θ *isomorphic to the symmetric group* S_3, *of order* 6 ; *the group* Θ *can be generated by the two following involutory collineations:*

$$\mu = [(x,y) \to (xy, y^{-1})] \quad , \quad \nu = [(x,y) \to (x^{-1}, xy)] \, .$$

Proof. The mappings μ and ν centralize every element out of $\Gamma_{(1,1)}$ and are collineations since the inverse property holds.

Since in Q holds $(xy)^{-1} = y^{-1} x^{-1}$ ([6], p. 111) one has $\mu\nu \neq 1$ and $(\mu\nu)^3 = 1$; therefore Θ has order 6 and acts on the three sets of horizontal, vertical and transversal lines as the symmetric group S_3. □

We are going now to study under which circumstances the direction preserving group Γ of collineations of a loop Q has some transitivity properties. The next theorem shows that the class of loops for which the collineation group Γ is transitive is very large.

Theorem (3.4). If Q *is a loop such that every element is a companion of a right and of a left pseudo-automorphism then the group* Γ *of collineations of* Q *which preserves the directions is point transitive, and viceversa .*

Proof. From the theorem (3.1) we know

$$\Gamma_{1_v} = \{(x,y) \to (x^\alpha, y^\alpha a)\}$$

$$\Gamma_{1_h} = \{(x,y) \to (b\,x^\beta, y^\beta)\}$$

where α is a right and β is a left pseudo-automorphism and a and b are companions of α and respectively of β .

The complex $\Phi = \Gamma_{1_v} \cdot \Gamma_{1_h} \subseteq \Gamma$ consists of the mappings $(x, y) \to (b\cdot x^{\alpha\beta}, (y^\alpha \cdot a)^\beta)$.

Therefore we have $(1,1)^\Phi = \{(b, a^\beta)\}$ where b and a are freely chosen in Q , and so the result follows. □

Theorem (3.5). For a loop Q *the following three conditions are equivalent:*

1) The direction preserving collineation group is point transitive on the net $N(Q)$.

2) Every element of Q *is a companion of a right and of a left pseudo-automorphism of* Q .

3) Every loop which is isotopic to Q *is isomorphic to* Q .

Proof. The equivalence of 1) and 2) is given by theorem (3.4). If Q' (with the identity $1'$) is isotopic to Q then Q' belongs to $N(Q)$ (but possibly $(1,1) \neq (1', 1')$). If there is a collineation β , mapping $(1,1)$ into $(1', 1')$ then Q is isomorphic to Q' and viceversa every isomorphism between Q and Q' induces a collineation in N(Q) (cf. [21], p. 50). □

The theorem (3.5) is a solution of the problem presented in [6], p. 57 as an unsolved problem: Find necessary and sufficient conditions for the loop Q in order that every loop isotopic to Q be isomorphic to Q . An algebraic expression for the required condition is that every element of Q is a companion of a right and of a left pseudo-automorphism.

Remark (3.6): Let Q be a loop such that every element of Q is a companion of a right (respectively left) pseudo-automorphism. If the collineation group of Q contains an element which interchanges the set of the horizontal lines and the set of the vertical lines then every element of Q is a companion also of a left (respectively right) pseudo-automorphism and the collineation group Γ of Q which preserves the directions is point transitive.

The geometrical theorems (3.1) and (3.4) can be applied to obtain algebraic results on loops with transitive automorphism group (cf. for definition [6], p. 88).

Theorem (3.7). If Q is a loop with a transitive automorphism group then exactly one of the three following properties occur:

1) *Only the identity is a companion of a pseudo-automorphism, i.e. every pseudo-automorphism is an automorphism,*
2) *Every element is a companion of a right (respectively left) pseudo-automorphism, but no element $\neq 1$ is a companion of a left (respectively right) pseudo-automorphism.*
3) *Every element of Q is a companion of a right and of a left pseudo-automorphism.*

Proof. Since every element different from 1 can be mapped by an automorphism on every other element different from 1 the stabilizer Γ_{1_v} respectively Γ_{1_h} is exactly then transitive on 1_v respectively 1_h when there exists an element different from 1 which is a companion. But if Γ_{1_v} respectively Γ_{1_h} is transitive then every element of Q is a companion. □

Theorem (3.8). If Q is a loop with a transitive automorphism group then either the right nucleus or the left nucleus consists only of the identity or every element of Q is a companion of a right and of a left pseudo-automorphism.

Proof. The assertion follows from the previous theorem if we remember that for every element c contained in the right nucleus respectively left nucleus the equation $(x\cdot y)c = x(y\cdot c)$ respectively $c(x\cdot y) = (c.x)y$ holds. □

Theorem (3.9). If Q is a proper commutative Moufang loop then the full collineation group of Q is not point transitive.

Proof. We consider the stabilizer Γ_{1_v} of the group Γ of all collineations which leave the set of vertical and the set of horizontal lines invariant. Since every pseudo-automorphism of Q is an automorphism of Q ([6], p. 115, thm 2.2) Γ_{1_v} cannot operate transitively on 1_v since the companions of automorphisms lie in the nucleus of Q which is a proper subgroup of Q ([6], p. 114, thm. 2.1). □

From (3.9) and (3.5) follows

Proposition (3.10) (cf. [1], [2] and [6], p. 58).
A necessary and sufficient condition that every loop isotopic to a Moufang loop Q be commutative is that Q be an abelian group.

Theorem (3.11). Let Q be a loop which possesses the inverse property and has a transitive automorphism group. Then either the left nucleus, the right nucleus and the middle nucleus consist only of the identity which is the only Moufang element of Q or Q is a proper non commutative Moufang loop in which every element is a companion of a pseudo-automorphism or Q is a group. If the nucleus of Q consists only of the identity then the collineation group of Q which preserves the directions consists exactly of the maps:

$$(x, y) \to (x^\alpha, y^\alpha)$$

where α is an automorphism of Q; also the full collineation group of Q leaves the point $(1,1)$ invariant.

Proof. Since Q has the inverse property, the left, middle, and the right nucleus coincide (cf. [6], thm. 2.1, p. 114). Now if the right nucleus is not equal 1 then the stabilizer Γ_{1_v} operates point transitively and every element of Q is a companion of a pseudo-automorphism of Q . Therefore Q is a Moufang loop ([6], lemma 2.2, p. 113). If Q is not a group then the first part of the assertions follows from theorems (3.8) and (3.9).

If β is a collineation of N(Q) which maps (1,1) on (m, n) then Q is isomorphic to the isotopic loop (Q, $*$) where the composition $*$ is given by

$$(x \cdot m) * (n \cdot y) = x \cdot y$$

with the identity n•m ([21], p. 48). Therefore the isotopic loop has also the inverse property and it follows that m and n are Moufang elements ([6], thm. 2.3, p. 115). But the only Moufang element of Q is the identity and the last assertion follows with (3.3). □

Corollary (3.12). Let Q be a connected topological loop which possesses the inverse property and which is realized on a 1-dimensional manifold. If Q has a transitive automorphism group then either the left nucleus, the right nucleus and the middle nucleus consist only of the identity which is the only Moufang element of Q or Q is a group (which is isomorphic to ($\mathbb{R}$, +) or to SO_2) .

Proof. From theorem (3.9) we have to exclude that Q is a proper non-commutative Moufang loop. This follows from [10], thm. (6.4) c) since Q is power associative. □

With the previous results we can also obtain some properties of division neorings (for the definition of this structure see [13],p. 507).

Theorem (3.13). If (R, +, .) *is a finite planar division neoring such that its multiplicative loop is power associative then the additive loop* (R, +) *is either an abelian group, or the left, the right and the middle nucleus of the additive loop* (R, +) *which must have the inverse pro-*

perty consist only of the neutral element: in particular (R, +) *has no Moufang elements* $s \neq 0$.

Proof. From Th(II, 8) in [13] we know that the loop (R, +) is commutative and has the inverse property. Now the assertion follows from thm. (3.9) and lemma 2.2 of [6], p. 113. □

From (3.13) we can deduce by help of (3.9) the

Remark (3.14). If (R, +) is a proper additive loop of a finite planar division neoring (R, +, .) whose multiplicative loop is power associative then the group Γ which preserves the directions leaves the point (1,1) fixed and operates transitively on the other points of a line through (1,1). Also the full group of collineations of (R, +) leaves (1,1) fixed.

Any division neoring F which has an element $a \neq 0$ contained in some additive subgroup possesses a prime subfield K lying in the center of F. The characteristic of K is called the characteristic of F (cf. [12] pp. 38-40).

A division neoring (R, +, .) is called a topological division neoring if the operations "+" and "." and all the binary operations which arise from solving the equations in (R, +) and (R, .) are simultaneously continuous in both variables.

Theorem (3. 15). Let (R, +, .) *be a division neoring with associative multiplication such that the additive loop* (R, +) *possesses the inverse property.*

(A). If the characteristic of R *is different from* 3 *then the neoring is either a skewfield or the left, the right and the middle nucleus of the additive loop consist only of the identity which is the only Moufang element of* (R, +).

(B). If (R, +, .) *is a connected, locally compact, finite dimensional, topological neoring then either* (R, +, .) *is one of the three classical fields (real numbers, complex numbers, quaternions) or the second alternative of (A) holds.*

Proof. From [12] , (1.18), p. 41 (cf. also [21] § 3.4 and [12] (4.13)) follows that (R, +) cannot be a proper Moufang loop. The rest follows from thm. (3.11) since (R, +) has a transitive automorphism group. □

We notice that there exist planar associative division neorings which possess an additive loop with the inverse property and which are homeomorphic to the real line $\mathbb{R}$ (cf. [22], pp. 459-461, § 13). The whole collineation group of such a loop leaves in the net N(Q) the point (1,1) fixed.

Paige gave necessary and sufficient conditions that the additive loop of an associative neoring is a commutative Moufang loop ([20] thm. II,11). From (3.11 (cf. also [5], 70 corollary 2)) follows that this condition can be satisfied only if the additive loop is an abelian group.

We study now the collineation group of a Moufang loop.

Theorem (3.16). Let Q *be a Moufang loop and* N(Q) *the net belonging to* Q *. Let us denote by* Σ *the full collineation group of* Q *and by* Γ *the subgroup of index* 6 *in* Σ *which leaves the set of horizontal lines and the set of vertical lines invariant.* Σ *(respectively* Γ *) operates transitively on the flags (respectively points) of* N(Q) *if and only if every element* $c \neq 1$ *of* Q *is a companion of a pseudo-automorphism of* Q *. On the set of lines* Σ *operates always transitively.*

Proof. The mappings

$$\gamma_a = [(x, y) \to (ax, ya); a \in Q] \qquad (o)$$

are collineations of N(Q) . For it is clear that γ_a maps the vertical (respectively horizontal) lines on vertical (respectively horizontal) lines. Let us consider now all points (x, y) with xy = d where d is a fixed element. Then we have ([6], p. 115, lemma 3.1)

$$(ax)(ya) = [a (xy)] a = ada$$

and so the image of a transversal line is a transversal line.

The set of the collineations (o) operates transitively on the set of the vertical (respectively horizontal) lines. The group Σ contains (see thm. (3.3)) an involution μ which maps the set of the transversal lines onto the set of vertical lines such that $(1_t)^\mu = 1_v$. Let now W be any transversal line; then there exist a γ_a with

$$(1_v)^{\gamma_a} = W^\mu = (1_t)^{\mu\gamma_a}$$

and we have $(1_t)^{\mu\gamma_a\mu} = W^{\mu\mu} = W$. Since Γ is a normal subgroup of Σ one has $\mu\gamma_a\mu \in \Gamma$ and Γ operates transitively also on the set of transversals.

Let us denote by Φ the subgroup of Γ generated by (o). Since $<\Phi\,\Gamma_{1_v}> \subseteq \Gamma$ we can map every point (x, y) on $(1,1)$ exactly then when every element of Q is a companion of a pseudo-automorphism of Q; using a suitable γ_a we obtain $(x, y)^{\gamma_a} = (1, t)$ and the rest of the statement follows from the theorem (3.4). □

§ 4. The Lenz classification for loops and 3-nets.

Let N be a k-net ($3 \leq k$). A translation α is a collineation of N which preserves the directions and leaves invariant every line of a direction $\mathcal{X}$. If $\alpha \neq 1$ we shall call the direction $\mathcal{X}$ the axis of α.

Remark (4.1). If $\alpha \neq 1$ is a translation of a k-net N, then α has no fixed points.

Proof. If s is a fixed point of α then every line incident with s consists only of fixed points. □

A collineation β of a k-net N ($3 \leq k$) which preserves the directions will be called a homology if all elements different from 1 of the group $<\beta>$ generated by β have exactly one fixed point p, the centre of β.

The translationswith the same axis $\mathcal{X}$ and all homologies with the same centre p form according to the case a subgroup $T(\mathcal{X})$ and $S(p)$ of the group Γ of all collineations which preserve the directions. We shall call the group $T(\mathcal{X})$ and then also the axis $\mathcal{X}$ transitive if the direction $\mathcal{X}$ contains a line G such that $T(\mathcal{X})$ is transitive on the points of G ; in this case $T(\mathcal{X})$ operates sharply point transitively on every line belonging to $\mathcal{X}$. In an analogous way we shall call the group $S(p)$ of homologies and then also the centre p transitive if $S(p)$ is transitive on the points, different from p, of a line G incident with p ; then the group $S(p)$ operates sharply transitively on the points different from p of every line incident with p.

Theorem 4.2. *If* N *is a 3-net then* N *belongs to exactly one of the following seven Lenz classes:*

I.1. - In N *there does exist neither a transitive axis nor a transitive centre.*

I.2. - In N *there is no transitive axis, but there exists exactly one transitive centre.*

I.3. - In N *there exists no transitive axis, but on every line there exists exactly one transitive centre.*

I.4. - In N *there exists no transitive axis, but the transitive centres of* N *are exactly the points of one line of* N .

I.5. - In N *there exists no transitive axis, but every point is a transitive centre.*

II.1 - In N *every direction is a transitive axis, but there is no transitive centre.*

II.2 - In N *every direction is a transitive axis and every point is a transitive centre.*

If Q is a loop then we will say that Q is of Lenz type A.a, where $A \in \{I, II\}$ and $a \in \{1,2,3,4,5\}$ if the net belonging to Q has the Lenz class A.a.

Let Q be a loop with respect to the multiplication " · " . The operations $(a, b) \to a\backslash b : Q \to Q$ respectively $(a, b) \to a/b : Q \to Q$ which are defined by $a\cdot(a\backslash b) = b$ respectively $(a/b)\, a = b$ give us on Q two further loop structures; one can assign in such a natural way to $(Q,\cdot)$ the right and the left reversed loop.

Under the cardinality of the isotopy class I(Q) of a loop Q we understand the number of different isomorphy classes of loops within I(Q).

Corollary (4.3). Every loop Q belongs to exactly one of the seven Lenz classes I.1 till II.2.

A loop Q is of type I.1 if and only if Q is not a group and no loop out of the isotopy class I(Q) admits a sharply transitive group of automorphisms.

A loop Q is of Lenz type I.2 if and only if the cardinality of the isotopy class I(Q) is at least five and I(Q) contains a loop admitting a sharply transitive group of automorphisms.

A loop Q is of Lenz type I.3 if and only if the cardinality of the isotopy class I(Q) is exactly two, I(Q) contains a loop Q^* admitting a sharply transitive group of automorphisms and no element $\neq 1$ of Q^* and its reversed loops is a companion of a pseudo-automorphism.

A loop Q is of Lenz type I.4 if and only if the cardinality of the isotopy class I(Q) is exactly two, I(Q) contains a loop Q^* admitting a sharply transitive group of automorphisms and every element of Q^* or of one of its reversed loops is a companion of a suitable right (respectively left) pseudo-automorphism.

A loop Q is of Lenz type I.5 if and only if Q admits a sharply transitive group of automorphisms, every loop isotopic to Q is isomorphic to Q and Q is not a group.

A loop Q is of Lenz type II.1 if and only if Q is a group which cannot be seen as the additive group of a vector space (over a field).

Q is of Lenz type II.2 if and only if Q is the additive group of a vector space (over a field).

<u>Remark (4.4).</u> If Q is a loop of Lenz type I.2 or I.3 which admits a sharply transitive group of automorphisms, then every pseudo-automorphism of Q is an automorphism.

This remark follows immediately from (4.3); if Q would admit a proper pseudo-automorphism α then every companion of α would be different from 1. Then, however, the stabilizer Γ_{1_h} or Γ_{1_v} of the group Γ of all collineations which preserve the directions would be transitive on the line 1_h or on the line 1_v (cf. 3.1). □

For the proof of (4.2) and (4.3) one uses the following

<u>Lemma (4.5).</u> Let Q be a loop and N(Q) the net belonging to Q; assume that in N(Q) there exists a transitive axis $\mathfrak{X}$. Then N(Q) belongs either to the Lenz class II.1 or to the Lenz class II.2. The net N(Q) belongs to the class II.1 exactly then if Q is a group

which cannot be seen as the additive group of a vector space; $N(Q)$ belongs exactly then to the class II.2 if Q is the additive group of a vector space (over a field).

Proof. From our hypothesis follows that every loop belonging to $N(Q)$ is a group isomorphic to Q and that every direction is a transitive axis ([4], p. 189). If $N(Q)$ does not belong to the class II.1, then every point p of $N(Q)$ is a transitive centre since the collineation group of $N(Q)$ is point transitive. Thus $N(Q)$ belongs to II.2. In this case every group Q which belongs to $N(Q)$ admits a sharply transitive group A of automorphisms.

The semidirect product $\theta = QA$ can be seen as a collineation group of $N(Q)$ which is contained in the stabilizer Γ_{1_h} of the group Γ of all collineations which preserve the directions, and which operates on 1_h sharply transitively. The nearfield F associated to θ has as additive group just Q and therefore Q is abelian ([16] (8.2)) . As a commutative group with a transitive automorphism group Q is the additive group of a vector space over a field ([6] , thm. 8.1). □

Proof of (4.2) and (4.3). Let N be a 3-net such that no direction is a transitive axis, but such that there exist two different transitive centres p_1 and p_2 .

We assume first that p_1 and p_2 are incident with a line L of the net N . Then the stabilizer Γ_L of the group Γ of all collineations which preserve the directions is point transitive on L . The line L can be seen as the line 1_h respectively 1_v of a loop Q such that either Q or one of the reversed loops of Q belongs to N . Since every element of Q is a companion of a left pseudo-automorphism respectively of a right pseudo-automorphism of Q , the group Γ operates point transitively on 1_v or 1_h (3.4). Thus N belongs either to the class I.4 or to the class I.5 according to the case whether there exists an element $\neq 1$ in Q which is companion of a left and of a right pseudo-automorphism or not.

If in N there is no line incident with the transitive centres p_1 and p_2, then the collineation group Γ is transitive on each one of the three sets of the horizontal lines, of the vertical lines and of the transversal lines. If N does not belong to the class I.5 then on every line of N there exists exactly one transitive centre. In this case N belongs to the class I.3.

If a loop Q is of Lenz type I.1 no loop in the isotopy class $I(Q)$ admits a sharply point transitive group of automorphisms (3.2).

If a loop Q is of Lenz type I.2 then the collineation group Γ of the net $N(Q)$ which preserves the directions, has at least five different orbits on the set of points. From [21] p. 50 (cf. also (3.5)) follows that the isotopy class $I(Q)$ has at least five different isomorphy classes of loops. From (3.2) is clear that the isotopy class $I(Q)$ contains a loop admitting a sharply transitive group of automorphisms.

If a loop Q is of Lenz type I.3 then the collineation group Γ of the net $N(Q)$, which preserves the directions, operates transitively on the points which are transitive centres. If we take a transitive centre as the point $(1,1)$ for a loop Q^* which belongs also to $N(Q)$ then Q^* admits a sharply transitive group of automorphisms (3.2). Therefore Γ is transitive on those points of N which are not transitive centres. Thus Γ has on N exactly two point orbits and the cardinality of the isotopy class $I(Q)$ is exactly two.

The rest of the assertions in (4.2) and (4.3) follows from (3.2), (4.5) and (3.5). □

Remark (4.6). Let Q be a loop which admits a sharply transitive group of collineations. Then the cardinality of the isotopy class $I(Q)$ is different from three and four.

Another characterization of loops of Lenz type I.4 is

Remark (4.7). A loop Q is of Lenz type I.4 if and only if the isotopy class $I(Q)$ contains a loop Q^* admitting a sharply transitive group of automorphisms and in Q^* or in one of its reversed loops there are elements $\neq 1$ which are companions of right (respectively left) pseudo-automorphisms but no elements $\neq 1$ which are companions of left (respectively right) pseudo-automorphisms.

The Lenz class I.1 contains not only loops satisfying only few algebraic rules (e.g. the free loops) but also for instance all proper Moufang loops admitting no transitive group of automorphisms. This follows e.g. from (4.4) and from the fact that in a Moufang loop there are always elements different from 1 which are companions of pseudo-automorphisms [6], p. 113 lemma 2.2 and [5], p. 70. Therefore every connected Lie Moufang loop is of Lenz type I.1 (cf. [17]). Also every finite, or every commutative, proper Moufang loop M is of Lenz type I.1; otherwise M would admit a sharply transitive group of automorphisms. Then M would be a simple loop such that no element has order 3 ([5], p. 70, cor. 2). If M is commutative we have a contradiction to (3.9) because of [6] p. 113 lemma 2.2 or p. 161 thm. 11.4. If M is finite then every element of M would be an involution ([9], p. 387) and this emplies again that M is commutative.

Also the Lenz class I.2 contains many examples of loops. For instance let (R, +, ·) be a division neoring with associative multiplication such that the additive loop (R, +) possesses the inverse property but is not a group. If either the characteristic of R is different from 3, or if R is a finite planar division neoring, or if R is a connected, locally compact, finite dimensional, topological neoring, then the loop (R, +) is of Lenz type I.2 (cf. (3.14) till (3.16), [12], [13], [22], pp. 459-461, [11] § 17, p. 229).

<u>Definition (4.8).</u> A k-net N is called <u>strongly planar</u> if it is embeddable in an affine plane E in such a way that the set of points of N and E is the same; moreover every translation of N can be extended to a collineation of E and every homology of N can be extended to a homology of E. If a k-net N is embedded in this way in an affine plane E, we say that N is <u>strongly embedded</u> in E.

There are many examples of loops of Lenz type I.1, I.2, II.1 and II.2. For instance strongly planar examples of groups of type II.1 can be constructructed in the following way. Let G be an infinite group which does not admit a sharply transitive group of automorphisms (e.g. let G be an infinite group in which there are elements of different order). From [14] and [25] follows that there exists a projective plane P with the following properties: In P there exists a point p on a line L such that the group A of elations with the centre p and the axis L is transitive and A is isomorphic to G. Consider now the affine plane P_L which arises from P by omitting the line L and all its points, and let N(p) be a 3-net consisting of 3 pencils of parallel lines of P_L, one of which is the pencil whose lines have the direction of the improper point p. A group which belongs to N(p) is strongly planar and of Lenz type II.1.

In contrast to the existence of many examples of the types mentioned above we have the following

Remark (4.9). There are no strongly planar 3-nets of Lenz types I.3, I.4, and I.5.

Proof. If N would be a 3-net of Lenz type I.3 or I.4 or I.5 which is strongly embedded in an affine plane A then it follows from [21] p. 67-70 that the collineation group of A would contain all translations of A. Then A would be desarguesian and N could be considered as a 3-net belonging to the additive group of a skew-field. But then N would belong to the Lenz class II.2. □

In general we have not been able to decide whether there exist examples of loops of Lenz type I.3, I.4 and I.5. If such examples exist, their order is at least 7 (cf. [8], § 4.2).

The loops of Lenz type I.4 are most peculiar. Since this class does contain neither commutative loops nor loops with the inverse property one cannot expect that the search for examples will be in the next time positive. Also our attempts to obtain examples of loops of Lenz type I.3 and I.5 in the class of additive loops of neofields ([20] and [15]) were not successful. On the other hand it is not known whether there exists a proper infinite simple Moufang loop admitting a sharply transitive group of automorphism; such a loop would be of the Lenz type I.5.

Another class of loops which may be considered in order to obtain examples is the class of totally symmetric loops. A totally symmetric loop is a commutative loop in which the following identity is satisfied: $x(xy) = y$. The totally symmetric loops correspond in a one-to-one way to the Steiner triple systems (cf. [8] p. 75); therefore there are totally symmetric loops which are not groups and which admit a sharply transitive group of automorphisms ([7],[18],[19],[23]) . Since the class of totally symmetric loops is not too difficult to handle we obtained the following

Remark (4.10). Let Q be a totally symmetric loop which admits a sharply transitive group of automorphisms. If Q possesses a pseudo-automorphism which is not an automorphism then Q is of Lenz type I.5.

Proof. Let N be the net belonging to the totally symmetric loop Q . The points of N are the pairs (x, y) with $x, y \in Q$ and the transversal lines c_t can be described by the equations $y = xc$. Let γ be a collineation of N which preserves the directions. Then γ can be described as a mapping of the type $(x, y) \to (x^\alpha, y^\beta)$ where α and β are permutations of Q such that for every $c \in Q$ there exists a suitable c' satisfying for all x the equation

$$x^\alpha (xc)^\beta = c' \tag{1}$$

For $x = 1$ from (1) follows $1^\alpha c^\beta = c'$ and we have

$$x^\alpha (xc)^\beta = 1^\alpha c^\beta \tag{2}$$

This equation for $x = c$ leads to $c^\beta = 1^\alpha(c^\alpha 1^\beta)$ and we have

$$x^\alpha (1^\alpha [(xc)^\alpha 1^\beta]) = c^\alpha 1^\beta \tag{3}$$

For $c = 1$ we have from (3):

$$x^\alpha (1^\alpha [x^\alpha 1^\beta]) = 1^\alpha 1^\beta = x^\alpha [x^\alpha (1^\alpha 1^\beta)]$$

and

$$1^\alpha (x^\alpha 1^\beta) = x^\alpha (1^\alpha 1^\beta) .$$

Now (3) is equivalent to

$$x^{\alpha}\,[(x\,c)^{\alpha}\,(\,1^{\alpha}\,1^{\beta}\,)\,] = c^{\alpha}\,1^{\beta} = x^{\alpha}\,[\,x^{\alpha}\,(c^{\alpha}\,1^{\beta}\,)\,] \qquad (4)$$

This emplies

$$(x\,c)^{\alpha}\,(1^{\alpha}\,1^{\beta}\,) = x^{\alpha}\,(c^{\alpha}\,1^{\beta}\,) \qquad (5)$$

If we take for α a proper pseudo-automorphism and for 1^{β} a companion of α then α leads to a collineation γ which does not leave the point (1,1) fixed. With our assumptions follows now that Q is of Lenz type I.5. □

In general we have been unable to decide whether or not there exist loops Q of Lenz type I.4 or I.5 admitting a group of collineations which in the corresponding net N(Q) preserves the directions, leaves a line L invariant and operates on the points of L sharply 2-transitively. The non-existence of such loops would follow from the non-existence of near-domains which are not near-fields.

Thus for instance there are no finite loops of such kind (cf. [16], p. 31) or no locally compact, connected loops with the above property (cf. [24]).

Literature

[1] J. Aczel, Quasigroups, nets and nomograms, Advances in Math. 1, (1965), 383-450.

[2] R. Baer, Nets and groups I, Trans. Amer. Math. Soc. 46 (1939), 110-141 .

[3] V.D. Belousov, Algebraic nets and quasigroups, (Russian). Izdat "Štiinca", Kišinev , 1971. 166 pp. MR 49 # 5214.

[4] W. Blaschke, Projektive Geometrie (3te Aufl.) Verlag Birkhäuser, Basel-Stuttgart 1954.

[5] R.H. Bruck, Some theorems on Moufang loops, Math. Z. 73 (1960), 59-78.

[6] R.H. Bruck, A survey of binary systems, Ergebnisse der Math. 20, Springer Verlag, Berlin-Heidelberg-New York (Third printing) 1971.

[7] M.J. Colbourn - R.A. Mathon, On cyclic Steiner 2-designs, Annals of Discrete Mathematics 7 (1980) 215-253, North-Holland Publ. Co. Amsterdam .

[8] J. Dénes - A.D. Keedwell, Latin squares and their applications, Akadémiai Kiado, Budapest 1974.

[9] S. Doro, Simple Moufang loops, Math.Proc. Cambr. Phil. Soc. 83, (1978), 377-392.

[10] K.H. Hofmann, Topologische Loops mit schwachen Assoziativitäts-forderungen, Math. Z. 70 (1958), 125-155.

[11] K.H. Hofmann, Topologische Doppelloops, Math. Z. 70 (1958) 213-230.

[12] K.H. Hofmann, Topologische distributive Doppelloops, Math. Z. 71 (1959), 36-68.

[13] D.R. Hughes, Planar division neo-rings, Trans. Amer. Math. Soc. 80 (1955), 502-527.

[14] D.R. Hughes, Additive and multiplicative loops of planar ternary rings, Proc. Amer. Math. Soc. 6 (1955), 973-980.

[15] A.D. Keedwell, On property D neofields, Rend. Mat. Pura e Appl., Roma 26 (1967), 383-402.

[16] W. Kerby, On infinite sharply multiply transitive groups, Hamburger Mathematische Einzelschriften, Vandenhoeck und Ruprecht, Göttingen, 1974.

[17] E.N. Kuz'min, Mal'cev, algebras and their representations, (Russian), Algebra i Logika 7 (1968), 48-69. Transl. Algebra and Logic 7 (1968), 233-244.

[18] E.H. Moore, Concerning regular triple systems, Bull. Amer. Math. Soc. 4 (1897), 11-16.

[19] E.H. Moore, Concerning abelian-regular transitive triple systems, Math. Ann. 50 (1898), 225-240.

[20] L.J. Paige, Neofields, Duke Math. J. 16 (1949), 49-60

[21] G. Pickert, Projektive Ebenen (Zweite Aufl.), Springer Verlag, Berlin - Heidelberg - New York, 1975.

[22] H. Salzmann, Topologische projektive Ebenen, Math. Z. 67 (1957), 436-466.

[23] P. Tannenbaum, Abelian Steiner triple systems, Canad. J. Math. 28 (1976), 1251-1268.

[24] J. Tits, Sur les groupes doublement transitifs continus, Comm. Math. Helv. 26 (1952), 203-224.
Corrections et complément, Comm. Math. Helv. 30 (1955) 234-240.

[25] P. Wilker, Doppelloops und Ternärkörper, Math. Ann. 159 (1965), 172-196.

ON POSSIBLE ORDERS OF NONCOMMUTATIVE TACTICAL SPACES

Johannes André
Fachbereich Mathematik
der Universität des Saarlandes
D-6600 Saarbrücken

Introduction

This paper will bring some remarks on non-commutative geometry (see e.g. [1,3,8,10]). Starting point is a skewaffine space (schiefaffiner Raum) being a space possessing a (generally noncommutative) join on every couple of points whose images will be point sets called lines (Linien); moreover an equivalence relation among these lines called parallelism is defined and some conditions hold in such a way that the space under consideration becomes an affine space if additionally the join is commutative. Important examples of skewaffine spaces are the group spaces belonging to certain permutation groups [1]. An affine space is a group space iff it is desarguesian. In the general case no purely geometric characterizations of group spaces among all skewaffine spaces are known (see, however, [4]).

A finite skewaffine space is tactical iff all its lines possess the same number of points; this number is called the order of this space. Any finite affine space is tactical and many results on possible orders, e.g. of nondesarguesian affine planes, are known ([6,7]).

In this paper some generalizations on noncommutative tactical spaces will be stated provided that the space under consideration possesses at least one straight line (commutative line, Gerade). The order is a prime power (Theorem 5.1) if such a space is a group space (and hence a space over a Frobenius group, see [2]). This is a generalization of the well known fact that a finite desarguesian plane has prime power order. The possible orders of tactical spaces with straight lines in which all parallelograms are closed coincide with that of finite affine planes (Theorem 7.1). Especially due to the well known results of Bruck and Ryser [6] there exist infinitely many numbers not occurring as orders in such spaces. No restrictions of the order, however, are known for arbitrary tactical spaces.

1. Skewaffine spaces.

A structure $S = (X, \sqcup, \|)$ with $X \neq \emptyset$, $\sqcup : X^2 \to PX$, $(x,y) \mapsto x \sqcup y$ and $\|$ an equivalence relation on $\sqcup(X^2) := \{x \sqcup y \mid x,y \in X\}$ is called a skewaffine space (in some earlier papers also called quasiaffine space, see e.g. [1,8]) if the following conditions hold:

(L0) $x \sqcup x = \{x\}$ for all $x \in X$,

(L1) $x,y \in x \sqcup y$ for all $x,y \in X$,

(L2) $z \in (x \sqcup y) \setminus \{x\}$ implies $x \sqcup y = x \sqcup z$,

(P0) $x \sqcup x \,\|\, y \sqcup y$ for all $x,y \in X$.

(P1) Given any $L = x \sqcup y$ and any $z \in X$ there exists exactly one $L' = z \sqcup w$ with $L \,\|\, L'$ (Euclidean condition).

(P2) $x \sqcup y \,\|\, x' \sqcup y'$ implies $y \sqcup x \,\|\, y' \sqcup x'$.

(TAM) For any $x,y,z,x',y' \in X$ with $x \sqcup y \,\|\, x' \sqcup y'$ there exists at least one $z' \in X$ with $x \sqcup z \,\|\, x' \sqcup z'$ and $y \sqcup z \,\|\, y' \sqcup z'$ (Tamaschke-condition).

Remarks. (1) The elements of X are called points. Point sets of the form $x \sqcup y$ with $x \neq y$ are called lines (Linien). Let $\mathcal{L}$ be the set of all lines. The point x is a basepoint (Aufpunkt) of the line $L = x \sqcup y$; this situation is denoted by $x \twoheadleftarrow L$.

(2) In some earlier publications (e.g. [2,3,5]) the conditions (L0) and (P0) are cancelled in the definition of a skewaffine space. The general theory of such spaces, however, does not change essentially by this slightly difference of the definitions.

(3) Lines L, L' with $L \,\|\, L'$ are called parallel. The line $L' \,\|\, L$ with $z \twoheadleftarrow L'$ uniquely determined by (P1) is denoted by

(1.1) $\qquad L' = (z \,\|\, L)$.

(4) A skewaffine space with commutative join $\sqcup$ is an affine space.

(5) Any line of a skewaffine space possesses either exactly one basepoint or all of its points are basepoints [3,5]. A line all of whose points are basepoints is called a straight line (Gerade). Let $\mathcal{G}$ be the set of all straight lines, then $\mathcal{L} \setminus \mathcal{G}$ is the set of all proper lines (i.e. lines with exactly one basepoint). A skewaffine space is affine iff all its lines are straight.

(6) If G is straight and $L \,\|\, G$ then also L is straight (Hereditary condition [3,5]).

(7) It is sometimes useful to describe the situation given in (TAM) more exactly by TAM (x,y,z,x',y'). Note that this expression makes sense if and only if $x \sqcup y \,\|\, x' \sqcup y'$.

2. Subspaces.

Let $S = (X, \sqcup, \|)$ be skewaffine. A subset $U \subseteq X$ is a <u>subspace</u> iff

(U) $x, y, z \in U$ <u>imply</u> $(x \| y \sqcup z) \subseteq U$.

This condition is equivalent with

(U1) $x, y \in U$ <u>imply</u> $x \sqcup y \subseteq U$

and

(U2) $x \in U$, $L \in \mathcal{L}$, $L \subseteq U$ <u>imply</u> $(x \| L) \subseteq U$

because $\|$ is an equivalence relation and (P1) holds. A skewaffine space only possessing the <u>trivial subspaces</u> $\emptyset$, $\{x\}$ and X is called <u>primitive</u>, otherwise it is <u>imprimitive</u>.

3. Group spaces.

Important examples of skewaffine spaces are the <u>group spaces</u>: Let G be a group acting transitively on X such that $G_x \setminus G_y \neq \emptyset$ for $x \neq y$ ($G_x \neq G_y$ suffices in the finite case); such a group is called <u>normally transitive</u> on X [1]. Define

(3.1) $x \sqcup y := \{x\} \cup G_x\, y$

and

(3.2) $L \| L'$ iff there exists a $g \in G$ with $L' = g\, L$.

The so defined space $V(G) := (X, \sqcup, \|)$ is skewaffine [1,3]. A purely geometric characterization of such spaces is not yet known. Obviously any $g \in G$ induces a <u>dilatation</u>, i.e. an automorphism mapping any line L into an $L' \| L$. The subspaces of a group space V(G) are exactly the blocks of G ([1], Satz 6.3), hence the space V(G) is primitive iff G is primitive (see e.g. [12], Sect. 8).

It may be $V(G) \cong V(G')$ but the groups G and G' are not isomorphic as permutation groups. Example: Both the group of all motions and the group of all proper motions on $\mathbb{R}^2$ generate the <u>circle-space</u> whose lines are the circles with their centers (these are the basepoints) and two circles are parallel iff they have the same radii. (A finite example due to P. Neumann may be found in [2], p.165.)

A property of a permutation group G is called <u>geometric</u> if $V(G) \cong V(G')$ implies that also G' has this property. Hence, primitivity and imprimitivity of groups are geometric properties, also the property to be an imprimitive Frobenius-group (cf. [2], but "primitive Frobenius-group" is no geometric property as P. Neumann has shown). It is an open problem whether the existence of a sharply transitive (normal) subgroup is a geometric property.

4. Tactical spaces.

A finite skewaffine space S is called tactical if $|L|$ is constant for all lines L of S. We call $k := |L|$ the order of such a space. All finite affine spaces are tactical; their orders coincide with those defined here. The case $k = 2$ implies $x \sqcup y = \{x,y\} = \{y,x\} = y \sqcup x$, i.e. all lines are straight. Therefore we always assume $k \geq 3$ henceforth.

THEOREM 4.1 (cf. [4], Satz 2.1). *A finite imprimitive skewaffine space is tactical iff the following two conditions hold* (such spaces are also called *semiaffine* [4]):

(S1) *For any subspace U and any line L either $L \subseteq U$ or $|L \cap U| = 1$ is true.*

(S2) *Let U be a subspace with $x,y \in U$ and $x \neq y$, let L and M be lines with $x \in L$, $y \in M$ and $L,M \nsubseteq U$, then $|L \cap M| \leq 1$.*

A finite transitive group G whose group space is tactical is called *3/2-fold transitive* ([12], p.24); it is normally transitive due to $k \geq 3$ (cf. [1], Satz 2.5). Hence, Theorem 4.1 and [2], Satz 3.3, imply

THEOREM 4.2 (Wielandt [12], Theorem 10.4). *A finite 3/2-fold transitive group is either primitive or a Frobenius-group.*

5. Orders of tactical group spaces with straight lines.

We state the question what integers k are orders of suitable tactical spaces. No restricting properties are known without any additional hypothesis. Therefore, we always assume (additionally to $k \geq 3$) that *the tactical space under consideration is imprimitive and possesses at least one straight line.*

THEOREM 5.1. *An imprimitive tactical group space with a straight line has prime power order.*

Proof. Let be $V(G) = (X, \sqcup, \|)$ a group space. Then G is a Frobenius-group on X because of Theorem 4.1 and [2], Satz 3.3, and hence normally transitive. Let H be a straight line. Due to [1], Satz 2.5, we have $|H| = k \geq 3$. Any $g \in G$ with $gx = y \neq x$ for $x,y \in H$ belongs to $G_{(H)} := \{g \in G \mid gH = H\}$ ([1], Satz 2.3). Thus $G_{(H)}$ is double transitive on H, it is even sharply double transitive as G is a Frobenius-group. A sharply double transitive permutation group has prime power order (cf. e.g. [12], Sect.11). □

Remark. This theorem is a generalization of the well known property that a desarguesian plane has prime power order. For a tactical desarguesian affine plane is a group space [4].

6. Parallelogram-closure-condition

We say the parallelogram-closure-condition (Parallelogrammschließungsbedingung) holds in a skewaffine space $S = (X, \sqcup, \|)$ iff the following condition is valid:

(PGM) For any $x,y,z \in X$ there exists at least one $w \in X$ with $x \sqcup y \| z \sqcup w$ and $x \sqcup z \| y \sqcup w$.

For the situation just described we also state PGM (x,y,z); it makes sense for arbitrary $x,y,z \in X$ (see also remark (7) in section 1). The condition PGM does not follow from the other axioms of a skewaffine space. For one can prove that in a space $S = (X, \sqcup, \|)$ with PGM the lattice $\mathcal{U}_x$ of all subspaces containing a fixed $x \in X$ is modular. (Hence it is possible to build up a dimension-theory in such spaces generalizing Tecklenburg's theory [10]; results about it will be published elsewhere.)

Example. Let be G a group with center $\{1\}$. The set of all mappings $x \mapsto axb$ $(x,a,b \in G)$ is a normally transitive group Γ acting on G. But PGM does not generally hold in the group space $V(\Gamma)$ because the lattice $\mathcal{U}_1$ is isomorphic to the subgroup-lattice of G which is not modular e.g. for finite simple non-abelian groups (cf. [11], especially p.13, Theorem 14).

THEOREM 6.1. The parallelogram-closure-condition PGM holds in any group space $V(G)$ of a normally transitive group G with a transitive abelian subgroup.

Proof. Let T be an abelian subgroup of G; it acts sharply transitively on X (e.g. [12], p.9, Prop.4.4). Given $x,y,z \in X$ there exist exactly one $t \in T$ and one $t' \in T$ with $tx = y$ and $t'x = z$ resp. Now it is easy to see

$$tt'x = t'tx \in (y \| x \sqcup z) \cap (z \| x \sqcup y) \ . \quad \square$$

7. Tactical spaces with straight lines and PGM.

Let $S = (X, \sqcup, \|)$ be a tactical skewaffine space of an order $k \geq 3$ with PGM and let H be a straight line on S. Let $x \in H$ be arbitrary but fixed and let L be a line $\neq H$ on S with basepoint x. Define

$$(7.1) \qquad P_{H,L} := P := \bigcup_{y \in H} (y \| L) \ .$$

Generally P is no subspace of S . But one has

(7.1') $$P = \bigcup_{z \in L} (z \,||\, H)$$

due to PGM . Moreover

(7.2) $$|P| = k^2$$

because of (S1) and (S2) . It is our aime to construct an affine plane of order k with P as its point set. In order to distinguish the lines of the space S from those of this plane we call its lines (to be defined below) A-lines. Also the points of the plane, i.e. the elements of P , are called A-points. The set $A_{H,L} = A$ is defined by

(7.3) $$A := \{H' \mid H \,||\, H' \subseteq P\} \cup \{M \in L \mid M \subseteq P \;,\; M \cap H \neq M\} \quad .$$

LEMMA 7.1. The incidence-structure $(P,A) =: \mathcal{A} =: \mathcal{A}_{H,L}$ defined by (7.1) and (7.3) is an affine plane of order k .

The proof will be given in several steps.

(i) Two different A-lines possess at most one common point. If both lines are parallel to H they have no common point. If only one or none of the two lines are parallel to H then the uniqueness follows from (S1) or (S2) resp. (cf. Theorem 4.1).

(ii) Two different A-points p,q lie on at least one A-line. Obviously $p \sqcup q$ is such a line if $p \sqcup q \,||\, H$. Assume $p \sqcup q \not\| H$. Due to (7.1) there exist $p',q' \in H$ with $p' \sqcup p, q' \sqcup q \,||\, L$ or $p = p'$ or $q = q'$. The proposition is true if $p' = p$ or $q = q'$ or $p' = q'$. Assume the contrary. Select $z \in H \setminus \{p',q'\}$; this is possible because of $k \geq 3$. Now TAM (z,p',p,z,q') and $p' \sqcup p \,||\, q' \sqcup q$ imply the existence of a $y \in q' \sqcup q$ with $z \sqcup p = z \sqcup y$. By (S2) this y is uniquely determined by z,p',p and q'. The mapping $z \mapsto y$ from $H \setminus \{p',q'\}$ into $q' \sqcup q$ thus defined is injective by (S2) . Neither q' nor $(p \,||\, H) \cap (q' \sqcup q) =: \{q^*\}$ (this point exists uniquely by PGM and (S2)) are image-points of that mapping. Hence it is bijective from $H \setminus \{p',q'\}$ onto $(q' \sqcup q) \setminus \{q',q^*\}$ because both sets contain k-2 points. Thus there exists a $z^* \in H \setminus \{p',q'\}$ with $z^* \sqcup p = z^* \sqcup y$ and this is an A-line containing p and q .

An immediate consequence of (i) and (ii) is

(iii) Two different A-points are contained in exactly one A-line.

(iv) Two different A-lines have empty intersection iff they are parallel as lines in S . This is obvious if both A-lines are parallel to H. Let be $H' \neq H$ an A-line parallel to H and $M \neq H$ an A-line with a basepoint x on H . Select $y \in M \setminus \{x\}$. Due to (7.1) there is a $y' \in H$ with

$y' \sqcup y \parallel L$. For $z^* \in H'$ there exists a $z \in H$ with $z \sqcup z^* \parallel L$. Application of TAM (y, y', x, z^*, z) implies the existence of an $x^* \in H$ with $y \sqcup x \parallel z^* \sqcup x^*$, whence $M = x \sqcup y \parallel x^* \sqcup z^*$ by (P2). Now PGM (x^*, x, z^*) implies $M \cap H' \neq \emptyset$.

Now let M, M' be A-lines $\neq H$ with different basepoints x, x' resp. on H. Assume first $M \parallel M'$. Then $y \in M \cap M'$ and (P2) would imply $y \sqcup x = y \sqcup x'$ contradicting (S1). Hence $M \cap M' = \emptyset$. Suppose now $M \nparallel M'$ and choose $y \in M \setminus \{x\}$. By results proved above we obtain $x'' \in H$ and $y' \in M' \cap (y \parallel H)$ with $x'' \sqcup y \parallel x' \sqcup y'$. Moreover $x \neq x''$ by $M \nparallel M'$. Now TAM (x, x'', y, x, x') implies $M \cap M' \neq \emptyset$. This completes the proof of (iv).

(v) <u>Given an A-point</u> p <u>and an A-line</u> M <u>with</u> $p \notin M$ <u>there exists exactly one A-line</u> M' <u>with</u> $p \in M'$ <u>and</u> $M \cap M' = \emptyset$. This is true for $M \parallel H$ by (iv). Let M be a line $\neq H$ with a basepoint $x \in H$. If $p \in H$ then $M' := (p \parallel M)$ is such a line. Let be $p \notin H$. Due to (iv) there is a $q \in (p \parallel H) \cap M$ and now PGM (q, x, p) and (P2) imply $x \sqcup q \parallel y \sqcup p$ for a suitable $y \in H$. Hence $(y \parallel M)$ is such a line. The uniqueness is always a consequence of (iv).

(vi) <u>Now it is an obvious consequence that</u> $A = (P, \mathcal{A})$ <u>is an affine plane of order</u> k. □

Lemma 7.1 implies that there corresponds to every tactical space with a straight line and PGM an affine plane of the same order. Conversely any finite affine plane is such a space. Hence

THEOREM 7.1. <u>The possible orders of tactical skewaffine spaces with at least one straight line and the parallelogram-closure-condition</u> PGM <u>coincide with the possible orders of finite affine planes</u>.

<u>Remark</u>. The proof of Lemma 7.1 shows that it suffices to require only PGM (x, y, z) for all x, y with $x \sqcup y \parallel H$ for a fixed straight line H instead of the general condition PGM.

<u>References</u>.

1. André, J.: Über geometrische Strukturen, die zu Permutationsgruppen gehören. Abh.Math.Sem.Univ.Hamburg <u>44</u>, 203-221 (1976)
2. André, J.: Zur Geometrie der Frobeniusgruppen. Math.Z. <u>154</u>, 159-168 (1977)
3. André, J.: Introduction to non-commutative affine geometry. Lecture Notes Kuwait-University. Kuwait 1979.
4. André, J.: Eine geometrische Kennzeichnung imprimitiver Frobenius-Gruppen. To be published in Abh.Math.Sem.Univ.Hamburg <u>51</u>.
5. André, J.: Nichtkommutative Geometrie und verallgemeinerte Hughes-Ebenen. To be published in Math.Z.

6. Bruck,R.H., Ryser,H.J.: The non-existence of certain finite projective planes. Canad.J.Math. 1, 88-93 (1949)

7. Dembowski,P.: Finite geometries. Berlin-Heidelberg-New York:Springer 1968

8. Hauptmann,W.: Kohärente Konfigurationen, quasiaffine Räume und distanzreguläre Graphen. Mitt.Math.Sem.Gießen 144, 1-83 (1980).

9. Huppert,B.: Endliche Gruppen I. Berlin-Heidelberg-New York:Springer 1967

10. Misfeld,J., Tecklenburg,H.: Dimension of nearaffine spaces. In: Proceed.Conf.Geometry and Diff.Geometry,Haifa. Springer-Verlag, Lecture Notes in Mathematics 792, 97-109 (1979)

11. Suzuki,M.: Structure of a group and the structure of its lattice of subgroups. Berlin-Göttingen-Heidelberg: Springer 1956

12. Wielandt,H.: Finite permutation groups. New York-London: Academic Press 1964

MATHIEU GROUPS, WITT DESIGNS, AND GOLAY CODES

by

Thomas Beth and Dieter Jungnickel

Così fan tutte

Amongst classical finite geometries the families of t-designs $S_\lambda(t,k;v)$ admitting a multiply transitive automorphism groups are of particular interest.
The most important examples of this type are given by the celebrated designs due to Witt, which possess the simple groups of Mathieu as groups of automorphisms.

A t-design $S_\lambda(t,k;v)$ (with $v \geq k \geq t \geq 1$ and $\lambda > 0$) is defined to be a system $\mathfrak{B}$ of blocks, which are k-subsets of a set of v points, such that for each t-subset T of X there are exactly λ blocks containing T. In the special case, when $\lambda = 1$, $(X,\mathfrak{B})$ is called a Steiner system. Till today only systems with $t \leq 5$ are known to exist if $t \neq k$. Except for the classical Witt designs S(5,6;12) and S(5,8;24)*) which will be discussed here, for only five other pairs (t=5;k) examples of such systems have been found - four of them are due to Denniston [9] and another one has been given by Mills [15] who has applied Denniston's methods. The system S(5,6;12) resp. S(5,8;24) were constructed by Witt using the Mathieu groups M_{12} resp. M_{24} (cf. [20]). He also sketched a uniqueness proof for these systems [21]. Alternative presentations were given by Carmichael [5], Lüneburg [13], Curtis [7], Cameron/Van Lint [5], Cameron [4] and Hughes [10]. Whereas all these authors either discuss only one of these structures or treat both designs seperately,

---

*) For the sake of brevity we denote an $S_1(t,k;v)$ by S(t,k;v).

in this article we want to present a method of construction, which simultaneously produces both 5-designs and their groups only using elementary combinatorial and group theoretic facts. Without further efforts this also allows the introduction of the Golay codes generated by the indicator functions of the blocks. The technique presented here has been developed in order to provide a proof of the existence and uniqueness of the Witt designs, avoiding the usual group-theoric computations needed for the construction (cf. Witt [20]) as well as escapeing the delicate geometric considerations of the uniqueness proofs (cf. Lüneburg [13]). Furthermore the presentation should provide enough details to be accessible to the reader without too much additional work.

The only occasion, where the reader has to perform lengthy, though elementary computations, arises in constructing an S(5,8;24) from PSL(2,23). By means of coding theory (cf. remark (9)) even this step can be avoided. Apart from this the proof technique is built up in a chain of steps which link both 5-designs, their groups and their codes. Moreover, this presentation enjoys the additional advantage of elucidating the close interrelations between these fields exemplarily.

Here we have to mention the sources we are especially obliged to; in the first place these are the papers [20], [21] by Witt who in these two papers full of new ideas in fact has created a program that even today has not been fully completed. Furthermore we are greatly indebted to D. R. Hughes, about whose construction (cf. [10]) of S(5,6;12) we were able to learn in one of his brilliant and stimulating lectures at Westfield College. Finally (!), we want to thank Professor Lenz who during the many discussions in preparation of our joint book [3] contributed to the simplicity and clarity of quite a few steps of this proof.

An even more detailed discussion of the ideas to be presented here (including several alternative ways of constructing the Witt designs and Mathieu groups) may be found in Ch. III of this book.

From this short introduction it should have become clear that there exists close relations between t-designs and t-fold transitive Permutations groups. The reader recalls the

(1) Observation

Let G be a t-fold homogeneous permutation group acting on the set X and let B be a subset of X where $t \le k = |B| < v = |X|$. Then $\mathfrak{D} = (X, B^G)$ is an $S_\lambda(t,k;v)$ with $b = |B^G| = |G|/|G_B|$ and $\lambda = |G|\binom{k}{t}/|G_B|\binom{v}{t}$. Here G_B denotes the setwise stabilizer of B in G.

Using this result (following Witt [21] and Carmichael [6]) we construct a Steiner System S(5,8;24) as follows:

(2) Theorem

Let $G = PSL(2,23)$ and $X = PG(1,23)$. Then $\mathcal{M}_{24} = (X, B^G)$ with $B = \{\infty,0,1,3,12,15,21,22\}$ is a Steiner System S(5,8;24).

Proof: As G operates 3-homogeneously on X, by (1) $\mathcal{M}_{24}$ is an $S_\lambda(3,8;24)$. Let U be the subgroup of G generated by $x \longmapsto \frac{1+x}{1-x}$ and $x \longmapsto \frac{3x+1}{x-3}$. Observing $B = \infty^U$ we conclude $b \le 3.23.11$ and thus $\lambda \le 21$. Simple, but lengthy computations show that there are exactly 21 blocks containing the set $\{\infty,0,1\}$ and that the derived structure $(\mathcal{M}_{24})_{\infty,0,1}$ is the projective plane of order 4. The assertion thus follows from the 3-homogeneity of G.

Another way of constructing an S(5,8;24) will be sketched in (9).

To begin with, we want to investigate the combinatorial properties of an arbitrary system S(5,8;24), the first of which is well known and can be derived by easy counting arguments. □

(3) Lemma

Let $\mathfrak{D}$ be an $S(5,8;24)$ and U be a subset of a block B in $\mathfrak{D}$. Then the number $n(B,u)$ of blocks C in $\mathfrak{D}$ with $B \cap C = U$ only depends on the parameter $u = |U|$. In fact, $n(B,u)$ thus may be denoted by n_u and takes the following values

$$n_8 = 1,\ n_7 = n_6 = n_5 = n_3 = n_1 = 0\ ,$$
$$n_4 = 4,\ n_2 = 16,\ n_o = 30.$$

(4) Lemma

Let $\mathfrak{D} = (X,\mathfrak{B})$ be an $S(5,8;24)$. For two disjoint blocks A and B in $\mathfrak{D}$, also $C = X \setminus (A \cup B)$ is a block in $\mathfrak{D}$.

Proof: Choose a 4-subset D of C and a point $x \in C \setminus D$. Let R be the block through $D \cup \{x\}$. If $R \neq C$, by (3) R meets one of A or B in two points containing another point x' of C. Let y and y' denote the two remaining points of C. The block S through $C \cup \{y\}$ contains also y' and meets one of A or B in two points. Now consider any block T through $\{x,y\}$ and a 3-subset of D. By (3) T must contain another point of C, thus meeting R or S in 5 points, a contradiction. □

(5) Lemma

Let $\mathfrak{D} = (X,\mathfrak{B})$ be an $S(5,8;24)$. For any two blocks A and B in $\mathfrak{D}$ with $|A \cap B| = 2$, their symmetric difference $D = A + B$ is a 12-subset, called a dodecad in $\mathfrak{D}$. There are at most 132 block pairs (Y,Z) with $D = X + Z$. In case of equality, the class $\mathfrak{E}_D$ of all blocks of $\mathfrak{D}$ intersecting D in exactly 6 points forms an $S(5,6;12)$ in D.

Proof: Any 5 points of D uniquely determine a block Y in $\mathfrak{D}$. Obviously the maximum number of partitions $D = Y + Z$ is obtained if any block Y determined by 5 points in D contains exactly 6 points of D such that $Z = D + Y$ is a block too. In this case there are $\binom{12}{5}/\binom{6}{5} = 132$ such partitions $D = Y+Z$, with the trivial consequence that $\mathfrak{E}_D$ forms an $S(5,6;12)$. □

(6) Theorem

Let $\mathfrak{D} = (X,\mathfrak{B})$ an $S(5,8;24)$ and $\mathcal{C}$ the subspace generated by $\mathfrak{B}$ in the GF(2)-vectorspace $(2^X,+)$. Then $\mathcal{C}$ has dimension 12 and consists of

(i) $\emptyset, X$;

(ii) the 759 blocks and their complements;

(iii) 2576 dodecads.

Furthermore $\mathcal{E}_D$ forms an $S(5,6;12)$ for each dodecad D in $\mathfrak{D}$.

Proof: Obviously, $\mathcal{C}$ contains the 759 blocks. Choosing two disjoint blocks A and B, by (4) $C = X + A + B$ is a block, too. Thus $\mathcal{C}$ contains X and the complements of blocks. As (by (3)) there are $759 \cdot \binom{8}{2} \cdot 16$ pairs of Blocks (A,B) with $|A \cap B| = 2$, by (5) there exist at least $759 \cdot 448/132 = 2576$ dodecads in $\mathcal{C}$. Observing that $\dim \mathcal{C} \leq 12$, as any two blocks intersect in an even number of points (cf. (3)), implying that $\mathcal{C}$ is contained in its dual, the assertions (i), (ii), (iii) follow immediately. □

(7) Corollary

For blocks A,B with $|A \cap B| = 4$, their symmetric difference is also a block. The complement of a dodecad is also a dodecad.

(8) Notation

Choose a fixed dodecad D in the Steiner System $\mathfrak{M}_{24}$ (as given in (2)) and denote the associated Steiner System $S(5,6;12)$ on D by $\mathfrak{M}_{12}$. In particular this implies the existence of an $S(5,6;12)$.

(9) Remark

The vectorspace $\mathcal{C}$ (as given in (6)) is known as the extended binary Golay code which is contained in any treatise of coding theory, e.g. Mc Williams/Sloane [14] or Van Lint [19]. $\mathcal{C}$ can also be introduced via

the group ring $R = GF(2)[\mathbb{Z}_{23}]$: The cyclotomic polynomial $\sum_{i=0}^{22} x^i$ splits over R into the factors $g(x) = x^{11} + x^9 + x^7 + x^6 + x^5 + x + 1$ and $g^*(x) = x^{11} + x^{10} + x^6 + x^5 + x^4 + x^2 + 1$; hence there are only 2^3 ideals in R, which are generated by the polynomials $x-1$, $g(x)$ and $g^*(x)$. The ideal $\mathcal{G}$ generated by $g(x)$ is a linear code with parameters $(23,12)$, whose parity check extension equals $\mathcal{C}$. Applying well known facts about Quadratic Residue Codes ($\mathcal{G}$ is such a code) one easily derives that the weights of the nonzero vectors are doubly even and at least 8. The 759 codewords of weight 8 are seen to form an $S(5,8;24)$ when observing that $\mathcal{G}$ is an 3-error-correcting perfect code.
Next we shall consider certain substructures of $\mathcal{M}_{12}$, based on the assertion that $\mathcal{M}_{12}$ is resolvable. In order to prove this we give a slightly stronger statement which will be used later. The proof is by counting.

(10) Lemma

Let $\mathcal{F}$ be an $S(5,6;12)$ and U be a subset of a block B in $\mathcal{F}$. Then the number $n(B,u)$ of blocks C in $\mathcal{F}$ with $B \cap C = U$ only depends on $u = |U|$ and hence may be denoted by n_u. Here $n_6 = 1$, $n_5 = n_1 = 0$, $n_4 = 3$, $n_3 = 2$, $n_2 = 3$ and $n_0 = 1$. In particular $\mathcal{F}$ has a unique resolution.

(11) Lemma

Let $\mathcal{D} = (V,\mathcal{B})$ be an $S(5,8;24)$, D a dodecad and D' its complement. Then there is a bijective correspondence between the pairs of points in D' and the pairs of parallel blocks in the Steiner system $\mathcal{E}_D$. Those blocks of $\mathcal{D}$ through a given point p of D' meeting D' in exactly 2 points form a Hadamard subdesign $S_2(3,6;12)$ of $\mathcal{E}_D$.

Proof: Any block B of $\mathcal{E}_D$ intersects D' in a 2-subset. The proofs of (5) and (6) show that $B + X$ is also a block meeting D' in the same

2-subset. B and $B+X$ form a parallel class of $\mathcal{E}_D$; thus the first assertion follows for reasons of cardinality. Let p be a point in D'; then 11 2-subsets $\{p,q\}$ with $q \in D' \setminus \{p\}$ induce 11 parallel classes in $\mathcal{E}_D$. No two non-parallel blocks (taken from these 11 classes) can meet in 4 points within D, as they also intersect in p; neither can they intersect in two points of D, thus they pass through exactly 3 points of D. By $\mathcal{H}$ we denote the structure induced on D by the 22 blocks in consideration. For any $x \in D$ any two blocks of $\mathcal{H}_x$ meet in two points and each block in $\mathcal{H}_x$ has 5 points. Following a well-known theorem of Ryser's (cf. [18], 8.2.2) $\mathcal{H}_x$ then is a symmetric $S_2(2,5;11)$ thus proving the second assertion. □

<u>(12) Corollary</u>

$\mathfrak{W}_{12}$ has exactly 12 Hadamard subdesigns $S_2(3,6;12)$.

<u>Proof</u>: By (11) there are at least 12 such subdesigns. Choosing two points p,q of $\mathfrak{W}_{12}$ each subdesign $S_2(3,6;12)$ contains two blocks A,B connecting p and q. Furthermore $A \setminus \{p,q\}$ and $B \setminus \{p,q\}$ are parallel lines of the affine plane $(\mathfrak{W}_{12})_{p,q}$ of order 3. In this plane there are only 12 pairs of parallel lines. □

Thus being motivated to study possible subdesigns $S_2(3,6;12)$ of an arbitrary $S(5,6;12)$ we want to prepare these investigations by giving two lemmas.

<u>(13) Lemma</u>

Let K,L,M,N be 4 distinct blocks of an arbitrary design $S_2(3,6;12)$ with $L \neq K'$ and $N \neq M'$ (Y' denoting the complement of a set Y). Then $K+L = M+N$ iff $\{K,L\}=\{M,N\}$ or $\{K,L\}=\{M',N'\}$.

<u>Proof</u>: We only have to prove sufficiency: Observing the implications

$$K+L = K+N \Rightarrow L = N$$

$$K+L = K'+N \Rightarrow L = N'$$

(as + is a GF(2)-operation) we assume $K+L=M+N$ and $M \neq K, K', L, L'$ and recall that any two blocks of an $S_2(3,6;12)$ intersect in 0 or 3 points (cf. Norman [16]). We define $A=K \cap L$, $B=K' \cap L$, $C=K \cap L'$ and $D=K' \cap L'$. As $\lambda_3=2$ we see that $M \cap L$ is distinct from A and B. W.l.o.g. we may assume that $M=\{a_1,a_2,b_3,c_3,d_1,d_2\}$ and $M'=\{a_3,b_1,b_2,c_1,c_2,d_3\}$ where $A=\{a_1,a_2,a_3\}$ etc. is defined in the obvious way. Our assumption then shows $M+N=(M \cap N') \cup (M' \cap N)=K+L=B \cup C$, thus giving the contradiction $M \cap N'=\{b_3,c_3\}$. □

<u>(14) Lemma</u>

Let $\mathfrak{F}=(X,\mathfrak{B})$ be an $S(5,6;12)$ and K,L be two blocks of $\mathfrak{F}$ with $|K \cap L|=3$. Then $K+L$ also is a block in $\mathfrak{F}$.

<u>Proof</u>: Let $A:=K \cap L$ and $\mathfrak{B}_A$ be the set of all 12 blocks passing through A. Then the derived structure $\mathfrak{F}_A=(X \setminus A, \{Y \setminus A \mid Y \in \mathfrak{B}_A\})$ is an affine plane of order 3, in which $D=K \setminus A$ and $B=L \setminus A$ form parallel lines. Thus $C=(X \setminus A) \setminus (B \cup D)$ is parallel to D and B in $\mathfrak{F}_A$ and therefore $G=A \cup C=(K+L)'$ (here ' again means complementation) in $\mathfrak{F}$. Since by (10) $\mathfrak{F}$ is resolvable the assertion follows. □

<u>(15) Theorem</u>

Let $\mathfrak{F}=(X,\mathfrak{B})$ be an $S(5,6;12)$ and $\mathcal{H}=(V,\mathfrak{A})$ a subdesign $S_2(3,6;12)$ of $\mathfrak{F}$. Then

$$\mathfrak{B}=\mathfrak{A} \cup \{A+B \mid A,B \in \mathfrak{A} \text{ and } A \neq B,B'\}.$$

Moreover, each permutation φ of X satisfying $\mathfrak{A}^{\varphi} \subseteq B$ induces an automorphism of $\mathfrak{F}$. In particular, each automorphism of $\mathcal{H}$ is an automorphism of $\mathfrak{F}$.

<u>Proof</u>: By (14) we see that besides the 22 blocks of $\mathcal{H}$, also the $\frac{22 \cdot 20}{4}$ blocks of form $A+B$ (occuring in 4 ways according to (13)) are contained in $\mathfrak{B}$. Since $\mathfrak{F}$ has exactly 132 blocks, the first assertion is

proved implying that on automorphisms. □

(15) was stimulated by a remark of Hughes. It provides motivation for showing that any S(5,6;12) contains an $S_2(3,6;12)$. The subsequent proof technique was first sketched in Beth [2]. First we agree on the following

<u>(16) Notation</u>

Let $\mathfrak{S} = (X,\mathfrak{B})$ be any S(5,6;12) and let K,L be two blocks with $|K \cap L| = 3$. Furthermore define the set A,B,C,D,G as in the proof of (14). Then A,B,C,D are four disjoint triads:

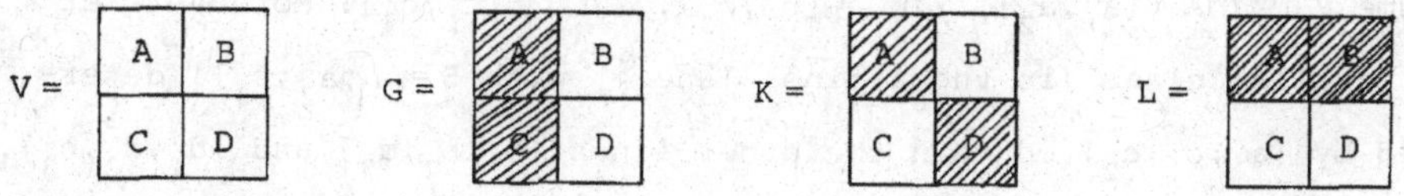

A subset X of V is said to be of type $\binom{\alpha\ \beta}{\gamma\ \delta}$ if $|Y \cap A| = \alpha$, $|Y \cap B| = \beta$, $|Y \cap C| = \gamma$, and $|Y \cap D| = \delta$.

<u>(17) Lemma</u>

Using the notation of (16) it follows that for any distinct points $a_1, a_2 \in A$, $c_1, c_2 \in C$, $b \in B$ and $d \in D$ there are exactly two blocks Y and Z of Type $\binom{2\ 1}{2\ 1}$ with $Y \cap A = \{a_1, a_2\}$, $Y \cap C = \{c_1, c_2\}$, $Z \cap B = \{b\}$ and $Z \cap D = \{d\}$.

<u>Proof</u>: By (10) there are exactly 3 blocks W with $W \cap G = \{a_1, a_2, c_1, c_2\}$, say W_1, W_2, W_3. Furthermore there exists a unique block V through $D \cup \{a_3, c_3\}$, where a_3 resp. c_3 are the third points of A resp. C. Since A and C are (parallel) lines of the affine plane $\mathfrak{S}_D$, it follows that V is of type $\binom{1\ 1}{1\ 3}$. Thus V' is of type $\binom{2\ 2}{2\ 0}$ containing $\{a_1, a_2, c_1, c_2\}$, say $V' = W_1$. Accordingly assume, that W_2 is of type $\binom{2\ 0}{2\ 2}$. Then the remaining block W_3 is of type $\binom{2\ 1}{2\ 1}$. This proves the first part of

the assertion. By the same reasoning there is exactly one block U of type $\binom{1\ 2}{1\ 2}$, containing $B \setminus \{b\}$ and $D \setminus \{d\}$. Then U' is of type $\binom{2\ 1}{2\ 1}$ containing b and d. The two other blocks S with $S \cap B = \{b\}$ and $S \cap B = \{d\}$ are of type $\binom{3\ 1}{1\ 1}$ resp. $\binom{1\ 1}{3\ 1}$. □

(18) Lemma

No two blocks of type $\binom{2\ 1}{2\ 1}$ intersect in exactly 4 points.

Proof: Let Y and Z be two blocks of type $\binom{2\ 1}{2\ 1}$ with $|Y \cap Z| = 4$. If $|Y \cap Z \cap A| = |Y \cap Z \cap C| = 1$, then also $|Y \cap Z \cap B| = |Y \cap Z \cap D| = 1$ contradicting (17). Again by (17), $|Y \cap Z \cap A| = |Y \cap Z \cap C| = 2$ is impossible. Thus we can assume $Y \cap Z \cap A = \{a_1, a_2\}$, $Y \cap C = \{c_2, c_3\}$, $Z \cap C = \{c_1, c_3\}$. Moreover let the parallel class (in the affine plane $\mathfrak{F}_E$ with $E = \{a_1, a_2, c_3\}$) determined by $\{a_3, c_1, c_2\}$ contain the other lines $\{b_1, b_2, d_3\}$ and $\{d_1, d_2, b_3\}$. The other lines of $\mathfrak{F}_E$ intersect the lines of this parallel class in a unique way. One of the blocks through a_1, a_2, c_1, c_3 is of type $\binom{2\ 2}{2\ 0}$ and another one of type $\binom{2\ 0}{2\ 2}$; w.l.o.g. we thus have that $\{c_1, d_2, d_3\}$ and $\{c_1, b_2, b_3\}$ are lines of $\mathfrak{F}_E$. Similary $\{c_2, d_1, d_3\}$ and $\{c_2, b_1, b_3\}$ are lines of $\mathfrak{F}_E$, which, being an affine plane of order 3, also contains the lines $\{c_1, b_1, d_1\}$, $\{c_2, d_2, b_2\}$ $\{a_3, d_3, b_3\}$, $\{a_3, b_1, d_2\}$ and $\{a_3, d_1, b_2\}$. The lines of $\mathfrak{F}_E$ thus are the rows, columns and transversals of the matrix

$$\begin{pmatrix} a_3 & c_1 & c_2 \\ d_3 & b_1 & b_2 \\ b_3 & d_1 & d_2 \end{pmatrix}$$

directly giving the equations $Y = \{a_1, a_2, c_2, c_3, b_2, d_2\}$ and $Z = \{a_1, a_2, c_1, c_3, b_1, d_1\}$. Thus $|Y \cap Z| = 3$ in contradiction to our assumption. □

(19) Lemma

No two blocks of type $\binom{2\ 1}{2\ 1}$ intersect in exactly 2 points.

<u>Proof</u>: By (17), there are 9 blocks of type $\binom{2\ 1}{2\ 1}$; consider the 36 pairs $\{Y,Z\}$ of such blocks and the corresponding 36 pairs $P_{YZ} = \{b_Y,d_Y\},\{b_Z,d_Z\}\}$ with $b_Y = B \cap Y$ etc. There are 18 pairs P_{YZ} with $b_Y \neq b_Z$ and $d_Y \neq d_Z$ and 18 pairs with $b_Y = b_Z$ or $d_Y = d_Z$. By (18), $|Y \cap Z \cap G| = 3$ implies $b_Y \neq b_Z$ and $d_Y \neq d_Z$; but there are 18 such pairs $\{Y,Z\}$ and thus the 18 pairs P_{YZ} with $b_Y = b_Z$ or $d_Y = d_Z$ correspond to the pairs $\{Y,Z\}$ with $|Y \cap Z \cap G| = 2$. □

<u>(20) Theorem</u>

Let $\mathfrak{S} = (X,\mathfrak{B})$ be an $S(5,6;12)$ and let K,L be two blocks of $\mathfrak{S}$ with $|K \cap L| = 3$. Then together with L,L',K,K' the blocks of types $\binom{2\ 1}{2\ 1}$ and $\binom{1\ 2}{1\ 2}$ form an $S_2(3,6;12)$ (using the notation of (16)). Thus any $S(5,6;12)$ contains a sub-design $S_2(3,6;12)$.

<u>Proof</u>: By (17), there are 9 blocks of type $\binom{2\ 1}{2\ 1}$ and similarly 9 blocks of type $\binom{1\ 2}{1\ 2}$. Denote the incidence structure formed by these blocks and L,L',K,K' by $\mathcal{H}$. By (18) and (19), any two blocks of $\mathcal{H}$ intersect in 0 or 3 points; one now sees that $\mathcal{H}$ is an $S_2(3,6;12)$ as in the proof of (11). Then the existence assertion is answered by (10). □

In order to show the uniqueness of the Steiner system $S(5,6;12)$ it will now suffice to show the uniqueness of $S_2(3,6;12)$ – in view of (15). This in turn is equivalent to showing the uniqueness of the derived Hadamard design $S_2(2,5;11)$ because of the following Lemma due to Alltop [1] and Norman [16]. The proof of this lemma is a standard exercise and will be left to the reader (or see [3]).

<u>(21) Lemma</u>

Let $\mathfrak{D}$ be an $S_\lambda(t,k;2k+1)$ where t is even. Then $\mathfrak{D}$ can be extended to an $S_\lambda(t+1,k+1;2k+2)$. In case of a Hadamard 2-design $S_\lambda(2,2\lambda+1\ ;4\lambda+3)$ this extension is unique (up to isomorphism) and affine resolvable.

We will also need some basic results on permutation groups which may be found in Huppert [12] or in [3] Ch. III, § 3.

(22) Lemma

a) Let G be a permutation group acting primitively on X and assume that the stabilizer G_x of some point x is simple. Then either G is simple or G contains a regular normal subgroup.

b) A nontrivial normal subgroup of a primitive group is transitive.

c) Let N be a transitive normal subgroup of a group G. Then G_x operates faithfully as an automorphism group of N.

d) Let N be a regular normal subgroup of a 2-transitive group G. Then N is elementary ablian.

We now sketch the uniqueness proof for $S_2(2,5;11)$ following Hussain [11]; like Hughes [10] we will simultaneously produce the automorphism group of this design.

(23) Theorem

There is a unique $S_2(2,5;11)$ which we will denote by $\mathcal{H}_{11}$. Aut $\mathcal{H}_{11}$ is a simple group of order 660 and acts 2-transitively on both the points and blocks of $\mathcal{H}_{11}$.

Proof: Let $\mathcal{L}$ be any $S_2(2,5;11)$ and choose any block $\mathfrak{B}$ of $\mathcal{L}$. Then the remaining blocks of $\mathcal{L}$ (by intersection) induce the complete graph K_5 on B; here the 5 blocks through any point $p \not I$ B induce a pentagon on B and any two of these 6 pentagons intersect in precisely 2 edges. It is easily seen that there is (up to isomorhism) only one solution for this problem:

The pentagon

P = 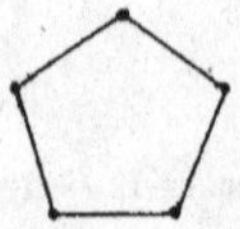

and the 5 images of

Q = 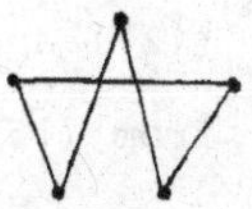

obtained from the 5 rotations which preserve P. This shows the existence and uniqueness of $S_2(2,5;11)$ if one checks that these 6 pentagons fulfil the required properties. It is easily observed that the design can be reconstructed uniquely from its "Hussain structure" determined by $\mathcal{L}$ and B. Of course the existence of some $S_2(2,5;11)$ is well-known and may also be obtained from the difference set {1,3,4,5,9} of quadratic residues in $\mathbb{Z}_{11}$ (the cyclic group of order 11). Using the uniqueness of the "Hussain structure" it is also easily seen that $G = \mathrm{Aut}\ \mathcal{H}_{11}$ is block-transitive and that the stabilizer G_B is isomorphic to A_5. Hence G has order 660 and acts 2-transitively on blocks (note that a block $C \neq B$ is uniquenely determined by the point pair $C \cap B$). Thus G is 2-transitive on points, cf. Dembowski [8] or [3], III, § 3. Any non-trivial normal subgroup N of G would be transitive by (22b) and would induce a normal subgroup isomorphic to $N \cap G_B$ of A_5; hence N would be regular and thus $A_5 \leq \mathrm{Aut}\ N \cong \mathrm{Aut}\ \mathbb{Z}_{11}$, a contradiction. □

(24) Corollary

There is a unique $S_2(3,6;12)$ which we will denote by $\mathcal{H}_{12}$. $\mathrm{Aut}\,\mathcal{H}_{12}$ is a simple group of order 7920 which acts 3-transitively on the points of $\mathcal{H}_{12}$.

This is an immediate consequence of (21), (23) and (22 a,d). Later we will show that $\mathrm{Aut}\ \mathcal{H}_{11}$ is isomorphic to PSL(2,11) and we will also determine $\mathrm{Aut}\ \mathcal{H}_{12}$. Now, let $\mathfrak{S}$ again denote any S(5,6;12). According to (20), $\mathfrak{S}$ has a Hadamard subdesign $S_2(3,6;12)\ \mathcal{H}$; also the blocks of $\mathcal{H}$ generate those of $\mathfrak{S}$ according to (15). Using (24), at once we obtain the uniqueness of S(5,6;12); thus $\mathfrak{W}_{12}$ (defined in (8)) is the only such Steiner system. Let $M_{12} = \mathrm{Aut}\ \mathfrak{W}_{12}$; because of (15), the stabilizer of a subdesign $S_2(3,6;12)$ of $\mathcal{W}_{12}$ in M_{12} is of order 7920. But there are precisely 12 such subdesigns (by (12)) and these subdesigns are isomorphic by (24). Then it is clear from (15) that M_{12} is transitive on these subdesigns and thus has order $7920 \cdot 12 = 12 \cdot 11 \cdot 10 \cdot 9 \cdot 8$. It is an elementary exercise to show that no nontrivial automorphism of an $S(5,6;12)$ can fix 5 (or more) points. This is left to the reader (or cf. [3], IV, § 1). Hence M_{12} is sharply 5-transitive on the points of $\mathcal{W}_{12}$. Next by $\mathcal{W}_{11}$ denote the derived structure at a point x of $\mathcal{W}_{12}$. Then $\mathcal{W}_{11}$ is an $S(4,5;11)$ with a sharply 4-transitive group $M_{11} = (M_{12})_x$. Note here that any automorphism of $\mathcal{W}_{11}$ extends to one of $\mathcal{W}_{12}$, as $\mathcal{W}_{12}$ is resolvable (by (10)) which shows that the extension of $\mathcal{W}_{11}$ to $\mathcal{W}_{12}$ is essentially obtained by complementation. This also implies the uniqueness of $S(4,5;11)$ using that of $S(5,6;12)$ and (21). Using (22) one shows that $\mathrm{Aut}\,\mathcal{W}_{12} = M_{12}$ is simple, as $\mathrm{Aut}\,\mathcal{H}_{12}$ is simple by (24) and is contained in M_{12} as the stabilizer of a sub-$S_2(3,6;12)$. Later we will prove likewise that M_{11} is simple by showing $M_{11} \cong \mathrm{Aut}\,\mathcal{H}_{12}$ (see (23)). We now have proved:

(25) Theorem

There is a unique Steiner system $S(5,6;12)$; its automorphism group M_{12} is simple and sharply 5-transitive. Similarly there is a unique $S(4,5;11)$; its automorphism group M_{11} is simple and sharply 4-transitive.

(26) Remark

The groups M_{11} and M_{12} are the little Mathieu groups. The stabilizer of a point in M_{11} is a sharply 3-transitive group isomorphic to the Twisted PGL(2,9); this group in turn is an extension of the affine group of the near field on 9 elements. The designs $\mathcal{W}_{11}$ and $\mathcal{W}_{12}$ are called the little Witt designs; $\mathcal{W}_{11}$ extends the Möbius plane $S(3,4;10)$ of order 3 which in turn is an extension of the affine plane S(2,3;9) of order 3.

In the sequel, we will use the large Witt design $\mathfrak{W}_{24}$ to obtain further properties of M_{12} and M_{12}; conversely the uniqueness of S(5,8;24) will be demonstrated using that of S(5,6;12). At this point, we first will show how one may use the uniqueness of S(5,6;12) to demonstrate that of the little Mathieu groups. We shall use the discussion of sharply t-transitive groups in Passman's book [17].

(27) Theorem

The only sharply t-transitive groups with $t \geq 4$ are the little Mathieu groups.

Proof: As in Theorem 21.5 of Passman [17] one proves the following result of Jordan: A sharply t-transitive group with $t \geq 4$ has either degree 11 (then $t = 4$) or degree 12 (then $t = 5$). Now let G be sharply 4-transitive of degree 11, say on $X = \{1,...,11\}$. According to the proof in [17] G has an elementary abelian subgroup H of order 4 fixing 7 and acting on $Y = \{8,9,10,11\}$ as the Klein Four group; furthermore $W = N_G(H)$ has order 24 and acts as S_4 on Y. Note that W fixes 7, as the normalizer $N_G(H) = W$ permutes the fixed points of H and as 7 is the only such point. Define $\alpha \in G$ by $8^\alpha = 8$, $9^\alpha = 9, 10^\alpha = 10$ and $11^\alpha = 7$; then $W^\alpha = (G_Y)^\alpha = G_{Y^\alpha}$ acts as S_4 on $Y^\alpha = \{7,8,9,10\}$ and fixes 7^α. Let $\beta \in G$ fix 8 and 11 and interchange 9 and 10; then $\beta \in W$ implies $7^\alpha = 7$, hence $\beta \in W^\alpha$. Thus β fixes 7^α; as β has at most 3 fixed points, one obtains $7^\alpha = 11$. Now let $B = Y \cup \{7\}$; as G_B contains both G_Y and α one sees that $G_B \cong S_5$. Now apply (1) with B as start block; this yields an S(4,5;11) with sharply 4-transitive automorphism group G; now (25) implies $G \cong M_{11}$. Next let $\bar{G}$ be sharply 5-transitive of degree 12; we may assume that $\bar{G}$ acts on $\{1,...,12\}$ and that $\bar{G}_{12} = G$.

Then $\bar{G}_B = G_B$ acts an S_5 on B and fixes 12. As in the previous case one shows that $\bar{G}_C$ with $C = B \cup \{12\}$ acts as S_6 on C; then (1) (applied to the start block C) yields an S(5,6;12) with the sharply 5-transitive automorphism group G. Again (25) implies $\bar{G} \cong M_{12}$. □

We next sketch (as in (9)) the connection between the little Witt designs and certain codes; i.e. the ternary Golay code.

(28) Remark

Constructing $\mathcal{H}_{11}$ from the difference set of quadratic residue on $\mathbb{Z}_{11}$ shows that ASL(1,11) is contained in Aut $\mathcal{H}_{11}$. But although $\mathcal{H}_{12}$ is the extension of $\mathcal{H}_{11}$ to the point set $\mathbb{Z}_{11} \cup \{\infty\}$, it is not true that PSL(2,11) (which is indeed an extension of ASL(1,11)) is contained (in its natural representation on GF(11) ∪ {∞}) in Aut $\mathcal{H}_{12}$. In fact $(GF(11) \cup \{\infty\}, \{\infty,1,3,4,5,9\}^{PSL(2,11)})$ is already an S(5,6;12), cf. Carmichael [6] or [3], IV, § 1; but we will see in (34) that Aut $\mathcal{H}_{11} \cong PSL(2,11)$. It can be shown that PSL(2,q) is an automorphism group for the Paley Hadamard design $\mathcal{H}_{q+1}$ ($q \equiv 3 \bmod 4$) iff $q = 7$, see Beth [2]. Nevertheless, the extended binary quadratic residue codes of length $q+1$ (with $q \equiv 7 \bmod 8$) admit PSL(2,q) as an automorphism group. But in our case $q = 11$ the binary code generated by the blocks of $\mathcal{H}_{12}$ is not too interesting: as 2 is a non-square in GF(11), one just obtains the hyperplane of words of even weight in $GF(2)^{12}$. On the other hand, any two blocks of $\mathcal{H}_{12}$ are orthogonal over GF(3), as shown in (13); also, the GF(2)-sums of non-parallel blocks generate an S(5,6;12) by (15). Thus it seems reasonable to consider the ternary code $\mathcal{T}$ generated by the blocks of $\mathcal{H}_{12}$. As Aut $\mathcal{H}_{11}$ contains the cyclic group $\mathbb{Z}_{11}$, we may consider the shortened code $\mathcal{T}_\infty \leq GF(3)^{11}$ (and generated by the blocks of $\mathcal{H}_{11}$) as an ideal in the group ring $R = GF(3)[\mathbb{Z}_{11}]$. Note that the polynomial $x^{11} - 1$ over GF(3) splits into the irreducible factors $x - 1$, $g(x) = x^5 - x^3 + x^2 - x - 1$ and $g^*(x) = x^5 + x^4 - x^3 + x^2 - 1$.

By our obversations above, $\mathfrak{T}$ is contained in its orthogonal complement $\mathfrak{T}^{\perp}$; also w.l.o.g. $\mathfrak{T}_{\infty}$ may be taken as the ideal generated by g in R. Then $\mathfrak{T}$ has dimension 6 and thus $\mathfrak{T} = \mathfrak{T}^{\perp}$. As 3 is a square in GF(11), $\mathfrak{T}_{\infty}$ is a quadratic residue code; the BCH-bound implies that $\mathfrak{T}_{\infty}$ has minimum weight ≥ 4. It may be shown that the weights in $\mathfrak{T}_{\infty}$ are all $\not\equiv 1 \bmod 3$; thus $\mathfrak{T}_{\infty}$ has in fact minimum weight ≥ 5. Then the sphere packing bound implies that $\mathfrak{T}_{\infty}$ is a perfect 2-error correcting code (the ternary Golay code). Conversely it would be possible to construct S(5,6;12) from $\mathfrak{T}$ defining blocks to be the supports of code words of weight 6.

Our next aim is the uniqueness proof for S(5,8;24) (using our result on $\mathfrak{M}_{12}$) and the construction of its group.

(29) Lemma

Let $\mathfrak{D} = (X,\mathfrak{B})$ be an S(5,8;24) and let D be a dodecad and D' its complement. Then the 132 blocks of $\mathfrak{B}_D$ (see (5)) generate all of $\mathcal{C}$ (see (6)), i.e. $\overline{\mathfrak{B}_D} = \mathcal{C}$.

Proof: Is suffices to show that $\overline{\mathfrak{B}_D}$ contains the 759 blocks of $\mathfrak{D}$. Let A be an arbitrary 4-subset of D and let $B_1, \ldots, B_4$ be the 4 blocks in $\mathfrak{B}_D$ containing A. Clearly the four 2-subsets $B_i \cap D'$ are pairwise disjoint; thus the fifth block B_5 of $\mathfrak{D}$ containing A is the union of the remaining 4 points of D' with A, i.e. $B_5 = X + B_1 + B_2 + B_3 + B_4$. This yields another $\binom{12}{4} = 495$ blocks in $\mathfrak{B}_D$ provided we can show $X \in \overline{\mathfrak{B}_D}$. To this purpose choose a point $p' \in D'$ and form the symmetric difference of the 11 blocks of the corresponding $S_2(3,6;12)$ (recall (11)) which pass through a point $p \in S$; as these 11 blocks form an $S_2(2,5;11)$ on $D \setminus \{p\}$ we indeed obtain X as their difference. It remains to show that each of the 132 blocks A with $|A \cap D| = 2$ and $|A \cap D'| = 6$ is in $\overline{\mathfrak{B}_D}$. Choose a 4-subset E of $A \cap D'$; then there is unique block C with $E \subseteq C$ and $|C \cap D| = |C \cap D'| = 4$ and we have already seen that $C \in \overline{\mathfrak{B}}$; hence also $A = (A + C) + C \in \mathfrak{B}_D$. ☐

(30) Theorem

Let $\mathfrak{D}_1 = (X_1, \mathfrak{B}_1)$ and $\mathfrak{D}_2 = (X_2, \mathfrak{B}_2)$ be Steiner systems $S(5,8;24)$ and let D_1 and D_2 be dodecads of $\mathfrak{D}_1$ and $\mathfrak{D}_2$ respectively. Furthermore denote the Steiner systems $S(5,6;12)$ induced on D_1 and D_2 (recall (6)) by $\mathcal{E}_1$ and $\mathcal{E}_2$ respectively. Then every isomorphism $\alpha : \mathcal{E}_1 \longrightarrow \mathcal{E}_2$ uniquely extends to an isomorphism $\bar{\alpha} : \mathfrak{D}_1 \longrightarrow \mathfrak{D}_2$.

Proof: For a point p in D_1' or D_2' let $\mathcal{H}(p)$ denote the subsystem $S_2(3,6;12)$ of $\mathfrak{D}_1$ resp. $\mathfrak{D}_2$ induced by p, see (11). Define β and γ by $p^\beta = \mathcal{H}(p)$ for $p \in D_1'$ and $q^\gamma = \mathcal{H}(q)$ for $q \in D_2'$. Now let B be any block of $\mathfrak{D}_1$ belonging to $\mathcal{E}_1$ and let $p \in D_1' \cap B$, i.e. $B \in p^\beta$. Then $B^\alpha \in p^{\beta\alpha}$ and if we want to extend α we have to define $\bar{\alpha}$ as follows:

$$p^{\bar{\alpha}} = \begin{cases} p^\alpha & \text{if } p \in D_1 \\ p^{\beta\alpha\gamma^{-1}} & \text{if } p \in D_1' ; \end{cases}$$

note that this is well-defined, as $p^{\beta\alpha}$ is a sub-$S_2(3,6;12)$ of $\mathcal{E}_2$ and as $\mathcal{E}_2$ contains exactly 12 such sub-designs by (12); so that $p^{\beta\alpha}$ indeed has a pre-image under γ. It is now easily seen that $\bar{\alpha}$ indeed maps the 132 blocks of $\mathfrak{D}_1$ belonging to $\mathcal{E}_1$ onto the 132 blocks of $\mathfrak{D}_2$ belonging to $\mathcal{E}_2$. In view of (29), $\bar{\alpha}$ is an isomorphism from $\mathfrak{D}_1$ onto $\mathfrak{D}_2$. ☐

(31) Theorem

There is a unique $S(5,8;24)$, say $\mathfrak{W}_{24}$. $\mathfrak{W}_{24}$ is resolvable and its automorphism group M_{24} is a 5-transitive group of order $24.23.22.21.20.48$. Forming stabilizers one obtains a 4-transitive group M_{23} and 3-transitive group M_{22}. These three groups are simple and are called the large Mathieu groups. The corresponding Steiner systems $\mathfrak{W}_{24}$, $\mathfrak{W}_{23}$ and $\mathfrak{W}_{22}$ are the large Witt designs. M_{12} is contained in M_{24} as the stabilizer of a dodecad.

Proof: The uniqueness of S(5,6;12) (see (25)) immediately implies that of S(5,8;24) by (30). Choose a dodecad D in $\mathfrak{M}_{24}$; from (30) we conclude $(M_{24})_D \cong \mathrm{Aut}\, S(5,6;12) = M_{12}$; again by (3), M_{24} is transitive on the 2576 dodecads. Thus $|M_{24}| = 2576.12.11.10.9.8 = 24.23.22.21.10.48$. As M_{12} acts (sharply) 5-transitively on the points of D, clearly M_{24} acts 5-transitively on points. In order to prove the resolvability of $\mathfrak{M}_{24}$, note that ASL(1,23) < PSL(2,23) < M_{24} by (2); now ASL(1,23) splits the block set of $\mathfrak{M}_{24}$ into 3 orbits of length 23.11 each. Choose a Singer group S (of order 12) in PSL(2,13) and let α be its element of order 3; S is normalized by a dihedral group H of order 24 and H acts regularly on PG(1,23) as is readily seen from Huppert [12] II.8.4 and II.8.5. Defining B as in (2), one therefore obtains a trio B, B^{α} and B^{α^2} of pairwise disjoint blocks. One now checks that ASL(1,23) produces a resolution of $\mathfrak{M}_{24}$ from this trio. It remains to prove the simplicity of the large Mathieu groups; here simplicity of M_{22} will yield that of M_{23} and M_{24} by using (22). Likewise, M_{22} will be simple if the stabilizer $(M_{22})_x$ of a further point x is simple. But $G = (M_{22})_x$ has order $21.10.48 = |PSL(3,4)|$ and is contained in PΓL(3,4) (as the triply derived structure of $\mathfrak{M}_{24}$ is the unique projective plane of order 4) as a subgroup of index 6. Hence $G \cap PSL(3,4)$ is a subgroup of index ≤ 6 in PSL(3,4); but PSL(3,4) is simple and one has $|PSL(3,4)| > 6!$ which implies that PSL(3,4) has no proper subgroup of index ≤ 6. Thus $G = PSL(3,4)$ is simple. □

Having just used M_{12} to construct M_{24}, we will now conversely use M_{24} to deduce the existence of an outer automorphism of M_{12}.

(32) Theorem

M_{12} admits an outer automorphism which switches two conjugacy classes of subgroups isomorphic to M_{11}. Considering M_{12} as Aut $\mathfrak{M}_{12}$, one of these conjugacy classes is the stabilizer of a point and the other one the stabilizer of a sub-$S_2(3,6;12)$ of $\mathfrak{M}_{12}$.

Proof: Let D be a dodecad in $\mathfrak{M}_{24}$ and D' its complement. By (31), $G = (M_{24})_D$ is isomorphic to M_{12} and there exists $\delta \in M_{24}$ switching D and D'. But $G^{\delta} = (M_{24})_{D'} = G$ and thus δ induces an automorphism of M_{12}. If this automorphism were inner then the stabilizer $G_{p'}$ of a point $p' \in D'$ would also be the stabilizer G_p of some point $p \in D$. But $G_{p'}$ stabilizes the sub-$S_2(3,6;12)$ $\mathfrak{H}_{12}$ of $\mathcal{E}_{D'}$ that corresponds to p' according to (11). But Aut $\mathfrak{H}_{12}$ has order $7920 = |M_{11}|$ by (24) and thus $G_{p'} = \text{Aut}\, \mathfrak{H}_{12}$; hence $G_{p'}$ is not the stabilizer of a point $p \in D$. □

This proof simultaneously yields the result on Aut $\mathfrak{H}_{12}$ already mentioned and thus shows the simplicity of M_{11}:

(33) Corollary

One has Aut $\mathfrak{H}_{12} \cong M_{11}$.

By arguments similar to that of (32) one may also construct an outer automorphism of S_6 using $\mathfrak{M}_{12}$ (as M_{12} acts on each block of $\mathfrak{M}_{12}$ as S_6); we will leave this to the reader. Next we shall show that Aut $\mathfrak{H}_{11} \cong PSL(2,11)$; this is also well-known and will be included for the sake of completeness as proofs are not that easily accessible in the literature.

(34) Theorem

One has Aut $\mathfrak{H}_{11} \cong PSL(2,11)$.

Proof: By the proof of (22) we know that $G = \text{Aut}\, \mathfrak{H}_{11}$ is a simple group of order $660 = 11.10.6$ containing A_5. Choose one of the 12 Sylow 11-subgroups of G, say P; then G acts on the 12 cosets of $N_G(P)$ by translation. As G is simple, this representation of G is faithful. One sees that P itself operates on the 12 cosets of $N_G(P)$ as follows: It fixes

$N_G(P)$ and permutes the other 11 cosets cyclically. Hence we may identify P with $\langle\tau\rangle$ on PG(1,11) where

$$x^{\tau} = \begin{cases} x+1 & \text{if } x \in GF(11) \\ \infty & \text{if } x = \infty \end{cases} ;$$

then $N_G(P)$ is a transitive permutation group on GF(11) and is thus (according to a theorem by Galois, see Huppert [12] II.3.6) permutation isomorphic to ASL(1,11), say $N_G(P) = \langle\tau,\pi\rangle$ where $x^{\pi} = 4x$. (This is not surprising: the difference set representation of $\mathcal{H}_{11}$ shows that $ASL(1,11) < G$). Then $\langle\pi\rangle$ is a Sylow-5-subgroup of G and one may assume $\pi \leq H \cong A_5$. But A_5 contains a dihedral group D_5 and thus G contains ρ with $\rho^2 = 1$ and $\rho\pi\rho = \pi^{-1}$. Now ρ has no fixed points as the stabilizer of any point in G has order 55; thus ρ interchanges the points 0 and ∞ which are fixed by π. Furthermore one has the condition

$$(*) \qquad (4x)^{\rho} = x^{\pi\rho} = x^{\rho\pi^{-1}} = x^{\rho}/4$$

for all $x \in GF(11)$. Let $1^{\rho} = c$; as ρ is a fixed-point-free involution, repeated application of (*) shows that we have $\rho : x \rightarrow c/x$ where c is a non-square in GF(11). Obviously τ, π and ρ generate PSL(2,11). □

Finally we give a proof that $A_8 \cong PSL(4,2)$ which also goes back to Witt [20].

(35) Theorem

One has $A_8 \cong PSL(4,2)$.

Proof: Let B be a block of $\mathfrak{M}_{24}$ and $G = M_{24}$. Then G_B acts 5-transitively on B and hence induces either A_8 or S_8 on B. But if G_B contained a transposition (on B), then PSL(2,4) would contain an element fixing 3 points of a line of PG(2,4) and interchanging the remaining two (note again that the three times derived structure of $\mathfrak{M}_{24}$ is the projective plane of order 4 and that the corresponding stabilizer of

3 points in M_{24} is PSL(3,4), cf. the proof of (31)). As this is not true, G_B induces A_8 in B; furthermore the pointwise stabilizer $G_{(B)}$ of B has order 16. Hence $G_{(B)}$ acts on each of the affine planes of order 4 determined by B as the translation group. Then $G_{(B)}$ is a regular normal subgroup of G_B (where G_B is considered as a permutation group on $X \setminus B$); thus $G_{B,x} \cong G_B/G_{(B)} \cong A_8$ (with x any point of $X \setminus B$) is a subgroup of the automorphism group of $G_{(B)}$ by (22c), i.e. A_8 is isomorphic to a subgroup of PSL(4,2), hence to PSL(4,2) for reasons of cardinality. □

Literatur

1. Alltop, W.D.: Extending t-Designs, J. Comb. Th. 18 (1975), 177-186

2. Beth, Th.: Some Remarks on Hughes' Construction of S(5,6;12). In: Finite Geometries and Designs, LMS Lecture notes 49 (1981) 22-30.

3. Beth, Th.; Jungnickel, D., Lenz, H.: Design Theory, to appear

4. Cameron, P.J.: Parallelisms of Complete Designs. LMS Lecture Notes 23, Cambrigde University Press (1976)

5. Cameron, P.J.; van Lint, J.H.: Graphs, Codes and Designs. LMS Lecture Notes 43, Cambridge University Press (1980)

6. Carmichael, R.D.: Introduction to the Theory of Groups of Finite Order, Boston (1937)

7. Curtis, R.T.: A New Combinatorial Approach to M_{24} Math. Proc. Cambridge Phil. Soc. 79 (1976), 25-41

8. Dembowski, P.: Verallgemeinerungen von Transitivitätsklassen endlicher projektiver Ebenen. Math. Z. 69 (1958), 59-89

9. Denniston, R.H.F.: Some New 5-Designs. Bull. London Math. Soc. 8 (1976), 263-267

10. Hughes, D.R.: A Combinatorial Cosntruction of the Small Mathieu Designs and Groups, to appear in Annals of Discrete Math.

11. Hussain, Q.M.: On the Totality of Solutions for the Symmetrical Incomplete Block Designs: $\lambda=2$, $k=5$, or 6, Sankhyā 7 (1945), 204-208

12. Huppert, B.: Endliche Gruppen I, Springer, Berlin-Heidelberg-New York (1967)

13. Lüneburg, H.: Transitive Erweiterungen endlicher Permutationsgruppen, Springer Lecture Notes 84 (1969)

14. McWilliams, F.J.; Sloane, N.J.A.: The Theory of Error-Correcting Codes, North Holland, Amsterdam-New York-Oxford (1978)

15. Mills, W.H.: A New 5-Design, Ars Combinatoria 6 (1980)

16. Norman, C.W.: A Characterization of the Mathieu Group M_{11} Math. Z. 106 (1968), 162-166

17. Passman, D.S.: Permutation Groups, Benjamin, New York (1968)

18. Ryser, H.J.: Combinatorial Mathematics, Wiley, New York (1963)

19. van Lint, H.J.: Coding Theory, Springer Lecture Notes 201 (1973)

20. Witt, E.: Die 5-fach transitiven Gruppen von Mathieu. Abh. Math. Sem. Hamburg 12 (1938), 256-264

21. Witt, E.: Über Steinersche Systeme, Abh. Math. Sem. Hamburg 12 (1938), 265-275

Y ya que las ideas
no son eternas como el mármol
sino immortales como una selva o un río,
la especulación anterior
asumió otra forma en el alba .....

Jorge Luis Borges.

Thomas Beth
Institut für Mathematische Maschinen
und Datenverarbeitung I der
Universität Erlangen-Nürnberg
Martensstr. 3
D-8520 Erlangen

Dieter Jungnickel
Mathematisches Institut der
Justus-Liebig-Universität
Gießen
Arndtstr. 2
D-6300 Gießen

Note added in proof: In a forthcoming paper, S.D. Smith has given a (somewhat related) approach to M_{24} (see S.D. Smith, Reconstructing M_{24} from its 2-local geometries).

EXTENDING STRONGLY RESOLVABLE DESIGNS

Albrecht Beutelspacher and Ursula Porta

Fachbereich Mathematik der Universität,
Saarstr. 21, D-6500 Mainz, W-Germany

A 2-(v,k,λ) design D with block set $\mathcal{B}$ is called *strongly resolvable*, if there exists a partition $\{\mathcal{B}_1,\ldots,\mathcal{B}_c\}$ of $\mathcal{B}$ in *classes* satisfying the following axioms:

(i) There exists an integer ρ such that through any point of D there are exactly ρ blocks of each class.

(ii) Any two distinct blocks of the same class intersect in a constant number μ_i of points.

(iii) Any two blocks of distinct classes have a constant number μ_o of points in common.

Obviously, strongly resolvable designs provide a common generalization of symmetric ($c = 1$) and affine ($\rho = 1$) designs. In this note we shall prove that any extendable strongly resolvable design is in fact symmetric or affine.

THEOREM. If D is an extendable strongly resolvable design, then D is symmetric or affine.

Since any extendable affine design is an affine plane (cf. for instance Dembowski[3], 2.2.20) and since the extendable symmetric designs were determined by Cameron[2], this Theorem characterizes (essentially) the extendable strongly resolvable designs.

1. Preliminary Results

Throughout this note, D denotes a strongly resolvable 2-(v,k,λ) design with b blocks, r blocks through any point, c classes, ρ blocks of each class through any point, m blocks in each class, "inner" constant μ_i, and "outer" constant μ_o.

RESULT 1 (Beker[1], Hughes and Piper[5]).

$$b = v+c-1.$$

If D is not symmetric, then $\mu_o = k^2/v$.

RESULT 2 (Harris[4]). D is affine if and only if k divides v.

2. Proof of the Theorem

Our Theorem will be proved by the following two Lemmas.

LEMMA 1. If D is extendable but not symmetric, then $k+1$ is a divisor of $b-r$.

Proof. It is well known (see for example [3], 2.2.16) that in any extendable design, $k+1$ divides $b(v+1)$. Hence $k+1$ is also a divisor of

$$b(v+1)\mu_o = \frac{vr}{k}(v+1)\frac{k^2}{v} = kr(v+1),$$

so $k+1$ divides $r(v+1)$.

On the other hand, $k+1$ divides

$$b(k+1) = bk + b = vr + b = r(v+1) - r + b.$$

Together, the assertion follows./

Remark. Denote by D^+ a smooth design in which any line has exactly $t+1$ points such that for a point p of D^+ the incidence structure $D^+(p)$ is a non-symmetric strongly resolvable design. Then by similar methods as in Lemma 1 one can prove that $kt+1$ is a divisor of $b-r$.

LEMMA 2. If D is not affine, then $k+1$ does not divide $b-r$.

Proof. Since D is supposed to be not affine, we infer from Result 2 that there exists a positive integer n with

(1) $$n < v/k = b/r < n+1.$$

Therefore

$$nr < b < (n+1)r,$$

so

$$n\rho = nr/c < b/c = m < (n+1)r/c = (n+1)\rho,$$

or

$$n\rho < m \leq (n+1)\rho - 1.$$

This yields

(2) $$nr < b \leq (n+1)r - c.$$

Our Theorem will be proved if we have shown the following two inequalities:

$$n-1 < (b-r)/(k+1) < n.$$

Using (2) we get on the one hand

$$\frac{b-r}{k+1} - (n-1) = \frac{b-r-(n-1)(k+1)}{k+1}$$

$$> \frac{nr-r-(n-1)(k+1)}{k+1} = \frac{(n-1)(r-k-1)}{k+1} \geq 0,$$

since in a non-symmetric design Fisher's inequality reads $r \geq k+1$.

On the other hand, (2), (1) and Result 1 imply

$$\frac{b-r}{k+1} - n = \frac{b-r - n(k+1)}{k+1} < \frac{(n+1)r-c - r - n(k+1)}{k+1}$$

$$= \frac{n(r-k-1) - c}{k+1} < \frac{v(r-k-1)/k - (b-v+1)}{k+1}$$

$$= \frac{v(r-k-1) - k(b-v+1)}{k(k+1)} = \frac{-v-k}{k(k+1)} < 0. \surd$$

References

1. Beker, H.: On Strong Tactical Decompositions. J. London Math. Soc. 16 (1977), 191-196.

2. Cameron, P.: Extending Symmetric Designs. J. Combinat. Theory (A) 14 (1973), 215-220.

3. Dembowski, P.: Finite Geometries. Berlin - Heidelberg - New York, Springer 1968.

4. Harris, R.: On Automorphisms and Resolutions of Designs. Ph. D. Thesis, University of London, 1974.

5. Hughes, D.R. and Piper, F.C.: On Resolutions and Bose's Theorem. Geom. Dedicata 5 (1976), 129-133.

SOME UNITALS ON 28 POINTS AND THEIR EMBEDDINGS

IN PROJECTIVE PLANES OF ORDER 9

A.E. Brouwer

Mathematisch Centrum
Kruislaan 413
Amsterdam

Abstract We answer three questions posed by F. Piper by exhibiting
(i) a unital that is not embeddable in a projective plane,
(ii) a unital which is embeddable, and isomorphic with its dual, but not the set of absolute points of a polarity (in fact examples exist in each of the four known projective planes of order 9),
(iii) a unital that can be embedded in two different planes.

Introduction

A unital is a 2-$(q^3+1,q+1,1)$ design. The classical unital with these parameters is obtained as the set of absolute points and nonabsolute lines of a unitary polarity of the projective plane $PG(2,q^2)$. Its full automorphism group is $P\Gamma U(3,q^2)$ (see O'NAN [8]). O'Nan proved that it does not contain a configuration of four blocks intersecting in six points; conversely one may ask whether any unital without "O'Nan configurations" is necessarily the classical one. This is probably true; apart from the trivial case $q = 2$ where the unique unital is the affine plane AG(2,3) (which indeed is free of O'Nan configurations) we have

Proposition A 2-(28,4,1) design without O'Nan configurations is the classical unital (for $q = 3$).

Proof. Exhaustive computer search. □

H.A. Wilbrink [11] characterizes the classical unital among the unitals by the absence of O'Nan configurations and some additional geometric condition.

F. Piper wrote in [9] a survey on unitals and posed several questions. To each of these questions the answer is 'no' - there exist hordes of 'ugly' unitals, counter-examples to anything you might conjecture. Even more is true: requiring that the unital be embeddable in a projective plane does not make it much nicer.

We constructed a large number (138) of 2-(28,4,1) designs; room considerations forbid us to list them explicitly so here only some statistics are given. (The designs with an automorphism of order 7 are listed in Brouwer [3]; printouts of the others are available upon request.) In the sequel 'unital' will often be synonymous with '2-(28,4,1) design'.

Given a design, one may consider the linear code generated by the rows of its incidence matrix (i.e., by the characteristic functions of the blocks) over some finite field GF(p). As is well known, this code is interesting only for p a prime dividing $r-\lambda$.

In our case $r-\lambda = 9-1 = 8$ so that we only need to consider binary codes. Note that both the code and its dual will contain the all-one vector $\underline{j}$ (the code, since the sum of all blocks is $r.\underline{j} = \underline{j}$, and its dual, since $k = 4$ is even). In the dual code only the weights 0,10,12,14,16,18,28 occur (all weights are even since the code contains $\underline{j}$, and a nonzero weight cannot be less than $r+1 = 10$) so that the weightenumerator $A(z) = \Sigma\, a_i z^i$ is determined completely by specifying a_{10}, a_{12} and a_{14}. The weightenumerator of the code itself then follows by the MacWilliams relations.
Now let us have a look at some individual units.

A. Two-transitive unitals

There are two unitals on 28 points with a two-transitive automorphism group: the classical unital and the Ree unital.

1. The classical unital

For an explicit description see [3]. This design is resolvable in 28 ways: for each tangent I at some point x of the unital U we find a resolution where the nine parallel classes are determined by the nine points of $I\backslash\{x\}$; each point y outside the unital is incident with 6 secants and 4 tangents, and the four points of intersection of the tangents with the unital form a block (namely $U \cap y^{\perp}$), so that y determines a set of 7 pairwise disjoint blocks. These are all the spreads (63 in total) and any set of 5 pairwise disjoint blocks is extendable to a spread. It is uniquely embeddable in a plane PG(2,9), and this is the desarguesian plane. It is isomorphic to its dual: $x \mapsto x^{\perp}$ defines an isomorphism. (The dual of an embedded unital is the structure consisting of the tangents and the exterior points; one verifies that this again is a unital.) Its automorphism group is $P\Gamma U(3,3^2)$ of order 12096; it is doubly transitive on the points of U. Each of its blocks is a Baer-subline (i.e., the intersection of a line and a Baer-subplane).

The code generated has dimension 21; for the weightenumerator see Andriamanalimanana [1]. It has $a_4 = 315$ so that we cannot retrieve the design as the words of minimal nonzero weight in the code. The dual code has weightenumerator $1 + 63z^{12} + 63z^{16} + z^{28}$ (It is easy to see that there cannot be words of weight 10 in the dual code - indeed, these would be conics in the projective plane entirely contained in the unital. But curves of degree 2 and 4 cannot have 10 points in common. Andriamanalimanana shows generally that when q is odd the classical unital in PG(2,q) never contains an oval. [Note however that there do exist unitals in the desarguesian plane containing ovals - we will see one below.])

The words of weight 12 in the dual code are unions of three blocks of the unital, such that the three lines carrying these blocks form a selfpolar triangle. Correspondingly we find 63 triples in the set of 63 exterior points, these being the lines of the well-known $G_2(2)$ generalized hexagon.

Any binary code with $n = 28$, $\dim = 7$, $d \geq 12$ and containing $\underline{j}$ must have this weightenumerator since it meets the Grey-Rankin bound with equality. Remains the question whether there is only one code with these parameters.

2. The Ree unital

For an explicit description see [3]. This design is resolvable in 10 ways, and has 45 spreads (any two resolutions having exactly one spread in common). There are no maximal partial spreads of size 6, but an embeddable unital must possess at least 63 partial spreads of size 6, so this unital is not embeddable in a projective plane. Its group is PΓL(2,8) of order 1512; it is doubly transitive on the points. In fact this is the smallest member in the family of Ree-unitals.

This design does contain O'Nan configurations, but has the property that each O'Nan configuration is contained in a set of five pairwise intersecting blocks, no three of which pass through the same point. It contains 126 of such super O'Nan configurations, 10 on each block.

The code generated has dimension 19. It has $a_4 = 63$, i.e., one gets the design back again by taking all words of weight 4. The dual code has weight enumerator

$$(1 + z^{28}) + 84.(z^{10} + z^{18}) + 63.(z^{12} + z^{16}) + 216.z^{14}.$$

(The group PGL(2,8) has index 3 in PΓL(2,8) and acts on the 28 points as a rank 4 group. This gives rise to a 3-class association scheme with $n_1 = n_2 = n_3 = 9$. The 84 words of weight 10 are the 'stars' in this association scheme: a point together with its i-th associates for i = 1,2 or 3.)

I conjecture that the Ree unital on 28 points is characterized by the fact that its code has dimension 19. (It is a well-known meta-conjecture that nice structures generate low-dimensional codes. In this sense the Ree unital is 'nicer' than the classical unital.)

Clearly, by the two-transitivity of the groups involved, all constant weight layers of the codes of these two unitals are 2-designs. Thus we find e.g. 2-(28,10,10), 2-(28,12,11) and 2-(28,14,52) designs.

B. Embeddable unitals

I found 11 unitals embeddable in a projective plane of order 9. All except one are uniquely embeddable, while the last one can be embedded in two ways - the resulting two planes are nonisomorphic: one is the translation plane and the other the dual translation plane. Below some statistics (s is the number of spreads, r the number of resolutions, aut the order of the automorphism group, dim the dimension of the code, a_{10}, a_{12}, a_{14} coefficients of the weight enumerator of the dual code, plane the projective plane containing the unital).

The design numbered E.0 seems to be the most popular one in the literature - almost all authors giving an explicit 2-(28,4,1) design in fact list this one - it has a short description in terms of $Z_3 \times Z_3 \times Z_3$ (see e.g. Hall [4]). Its pointset is the union of three ovals that are mutually tangent in one point ∞. From a theorem of Lefevre-Percsy [6] ("The Buekenhout-Metz unitals in a desarguesian plane of order $q^2 > 4$ are exactly those such that for some tangent line I to the unital U all Baer-sublines that meet I intersect U in 0,1,2 or q+1 points.") it follows that we have a Buekenhout-Metz unital (- here the tangent I is the tangent at ∞). Its dual code is not self-orthogonal, while it in fact is in the remaining ten cases.

Of course E.9 is the classical unital.

Unital	dual	s	r	aut	dim	a_{10}	a_{12}	a_{14}	plane
E.0	self	9	1	216	25	3	0	0	Des.
E.1	E.2	1	0	24	26	0	0	2	dual tr.
E.2	E.1	1	0	24	26	0	0	2	tr.
E.3	E.4	15	1	48	25	0	3	0	dual tr.
E.4	E.3	15	1	48	25	0	3	0	tr.
E.5	E.6	4	0	6	26	0	0	2	Hughes
E.6	E.5	4	0	6	26	0	0	2	Hughes
E.7	self	3	0	3	26	0	0	2	Hughes
E.8	self	7	0	48	24	0	3	8	Hughes
E.9	self	63	28	12096	21	0	63	0	Des.
E.10	self	31	0	192	22	12	15	8	tr./dual tr.

(Note that a unital and its dual must have the same group of automorphisms when the unital is uniquely embeddable in a projective plane.)

I believe that at least the unitals in the Hughes plane are new. (Except for the classical one, none of these unitals is derived from a polarity; the translation plane and dual translation plane do not have a polarity, and Piper [10] showed that the Hughes plane does not possess a unitary polarity.)

C. Miscellaneous

It seems clear that the combinatorial explosion in the number of nonisomorphic Steiner systems S(2,4,v) occurs at v = 28. For v ≤ 16 the system is unique, and for v = 25 I can construct exactly 4 of them (with automorphism groups of order 504, 150, 63 and 21); it seems difficult to construct others. For v = 28 however, I produced 138 nonisomorphic ones with very little effort; I am sure that the actual number of nonisomorphic solutions is much larger than 10^3.

Let us list some more statistics. The unitals found had automorphism groups of the following order:

Order	freq.	Order	freq.
1	26	21	4=
2	19	24	5
3	6	32	2
4	27	42	1=
6	7	48	10
7	1=	64	1
8	3	192	2
9	14	216	1
12	2	1512	1=
16	5	12096	1=

All unitals with an automorphism of order 7, 9 or 12 were determined. In particular the frequencies listed for orders divisibl by 7 are the actual frequencies. In the remaining orders only primefactors 2 and occur.

Five of the unitals found are resolvable (they were all mentioned above: the Ree unital and four embeddable unitals).
Five of the unitals found contain maximal partial spreads of size three. If s_i denotes the number of maximal partial spreads of size i (and s_7 := s) we have

Unital	s_3	s_6	s_7	r	aut	dim	a_{10}	a_{12}	a_{14}
E.0	72	54	9	1	216	25	3	0	0
C9.8	3	30	0	0	9	27	0	0	0
C9.12	3	48	0	0	9	27	0	0	0
3.22	1	36	3	0	3	27	0	0	0
C12.9	48	88	11	0	48	24	0	3	8

Twenty-two different weightenumerators occurred. The table below gives the dimension of the code and some parameters of the dual code. Note that a_{12} always equals a power of 2 minus one. Why?(*) Note that the classical unital is not alone in having a code of dimension 21; it is the only one of that dimension without words of weight 10 in the dual code.

Dim	a_{10}	a_{12}	a_{14}	selforthogonal	Dim	a_{10}	a_{12}	a_{14}	selforthogonal
19	84	63	216	no	23	8	7	0	yes
21	0	63	0	yes	24	0	3	8	yes
21	20	31	24	yes	24	0	7	0	yes
21	24	15	48	no	24	4	3	0	yes
22	8	15	16	yes	25	0	3	0	yes
22	12	15	8	yes	25	2	1	0	yes
23	0	3	24	no	25	3	0	0	no
23	0	15	0	yes	26	0	0	2	yes
23	4	3	16	no	26	0	1	0	yes
23	4	7	8	yes	26	1	0	0	yes
23	8	3	8	no	27	0	0	0	yes

References

[1] Bruno Ratsimandefitra Andriamanalimanana, *Ovals, unitals and codes*, dissertation Lehigh University, 1979.

[2] F. Buekenhout, *Existence of unitals in finite translation planes of order* q^2 *with a kernel of order* q, Geometriae Dedicata 5 (1976) 189-194.

[3] A.E. Brouwer, *Some unitals on 28 points and their embeddings in projective planes of order 9*, Math. Centre report ZW155, Amsterdam, March 1981.

[4] Marshall Hall jr., *Combinatorial theory*, Blaisdell-Wiley, 1967.

[5] W.M. Kantor, *2-Transitive designs*, in: *Combinatorics*, M. Hall & J.H. van Lint (eds.), Math. Centre Tracts 57, Amsterdam, 1974, pp. 44-97.

[6] C. Lefevre-Percsy, *Characterization of Buekenhout-Metz unitals*, preprint.

[7] R. Metz, *On a class of unitals*, Geometriae Dedicata 8 (1979) 125-126.

[8] M.E. O'Nan, *Automorphisms of unitary block designs*, J. Algebra 20 (1972) 495-511.

[9] F. Piper, *Unitary block designs*, in: *Graph Theory and Combinatorics* (R.J. Wilson, ed.), Research Notes in Math. 34.

[10] F. Piper, *Polarities in the Hughes plane*, Bull. London Math. Soc. 2 (1970) 209-213.

[11] H.A. Wilbrink, *A characterization of the classical unitals*, Math. Centre report ZW157, Amsterdam, March 1981.

(*) cf. Note p. 188

Note

It is easy to answer the 'Why?' on the previous page. In fact a word of weight 12 in the dual code is the sum of three blocks (since $r = 9$ and each block meets a given word w of weight 12 evenly, each point of w is on a unique block contained within w) and therefore also in the code itself; consequently words of weight 12 in the dual code meet in an even number of places, and the words of weight 0,12,16,28 form a linear subspace of the dual code. Hence $a_{12} = a_{16} = 2^j - 1$ for some j.

Having observed this we can also answer another question raised earlier in the paper. The dual code for the classical unital is not characterized by its weightenumerator $1 + 63z^{12} + 63z^{16} + z^{28}$ since we can extract from the dual code for the Ree unital a code with the same parameters (and it is impossible that both $P\Gamma L(2,8)$ and $P\Gamma U(3,3^2)$ act on the same code).

Amsterdam, 810524

THE LARGE WITT DESIGN - MATERIALIZED

Walter Fumy

Abstract And Introduction

The close connections between the Witt designs S(5,8;24), S(5,6;12) and the Mathieu groups $\mathcal{M}_{24}$ and $\mathcal{M}_{12}$ have been researched thoroughly and are commonly known [7], [5], [1]. A new derivation of the structures, giving the S(5,8;24) and the S(5,6;12) at the same time, combined with the presentation of their automorphism groups $\mathcal{M}_{24}$ and $\mathcal{M}_{12}$, respectively, was presented by T. Beth and D. Jungnickel [1]. As the authors have pointed out, the operation of finding the unique block that contains five given points can be performed for both Steiner systems using methods of coding theory [3], [5], [6]. As the blocks of the large Witt design generate the extended binary Golay code, the above operation can be treated as part of the decoding problem of this code. There exist a great number of wellknown decoding algorithms for the (24,12) Golay code [3]. Two of these decoding procedures make use of the properties of the large Witt design. The method of I.B. Gibson and I.F. Blake [4] is based on R.T. Curtis' miracle octad generator [2]. The more elegant method of J.M. Goethals [5] is a clever threshold decoding scheme and will be presented in this paper (section 2). Section 1 briefly discusses some properties of the extended binary Golay code. The final section will contain some remarks on the implementation of the decoding procedure.

1 The Extended Binary Golay Code

The extended binary Golay code G_{24} is a linear code of length 24, generated by the blocks of the Steiner system S(5,8;24), i.e., G_{24} is the GF(2)-rowspace of the incidence matrix of the large Witt design. The properties of this code can be derived from the properties of the S(5,8;24) on which the proposed decoding principles will be based.

Let $\mathfrak{D} = (X, \mathfrak{B})$ be an S(5,8;24).

(1.1) Lemma: Let b_i denote the number of blocks $B \in \mathcal{B}$ containing a chosen i-subset of X. Then $b_o=759$, $b_1=253$, $b_2=77$, $b_3=21$, $b_4=5$, and $b_5=\lambda=1$ holds.

(1.2) Lemma: Let U be a subset of a given block B in $\mathcal{D}$. Then the number n(B,U) of blocks C in $\mathcal{D}$ with $B \cap C=U$ depends only on $u=|U|$ and may thus be denoted by n_u, having the following values

$$n_8=1,\ n_7=n_6=n_5=n_3=n_1=o,\ n_4=4,\ n_2=16, \text{ and } n_o=3o.$$

Proof: cf. [1,(3)].

(1.3) Lemma: Let A,B be blocks in $\mathcal{D}$ with $|A+B|=12$ (here '+' denotes the symmetric difference). Then the set A+B is called a dodecad. There are at most 132 pairs of blocks (Y,Z) fulfilling A+B=Y+Z.

Proof: cf. [1,(5)].

(1.4) Definition: Let $n \in \mathbb{N}$, $k \in \mathbb{N}_n$. A binary linear (n,k)-code $\mathcal{C}$ is a k-dimensional linear subspace $\mathcal{C}$ of $GF(2)^n$. The linear space $\mathcal{C}^\perp$ orthogonal to $\mathcal{C}$ is called the dual code of $\mathcal{C}$.

(1.5) Definition: Let $n \in \mathbb{N}$. For a vector $\underline{x}=(x_1,x_2,\dots,x_n) \in GF(2)^n$ the set $supp(\underline{x}) = \{ i \in \mathbb{N}_n \mid x_i = 1 \}$ is called the support of $\underline{x}$. The cardinality $wgt(\underline{x}) = |supp(\underline{x})|$ is called the weight of $\underline{x}$. A binary linear (n,k)-code $\mathcal{C}$ is said to be e-error-correcting, if $\min\limits_{\substack{c \in \mathcal{C} \\ c \neq o}} wgt(\underline{c}) \geq 2e+1$.

Let G_{24} be the row space of the incidence matrix of $\mathcal{D}$.

(1.6) Observation: G_{24} is a binary linear code of length 24 containing
(i) 1 vector of weight o,
(ii) at least 759 vectors of weight 8,
(iii) at least 2576 vectors of weight 12,
(iv) at least 759 vectors of weight 16, and
(v) 1 vector of weight 24.

Proof: (i) and (ii) are consequences of the construction. Taking the sum of all incidence vectors $\underline{1}_B$, $B \in \mathcal{B}$, you get the vector of weight 24 since $b_1=253$ is odd. The vectors of weight 16 are the complements of the vectors of weight 8. If two blocks $A,B \in \mathcal{B}$ intersect in exactly two points, then $wgt(\underline{1}_A,\underline{1}_B)=12$ and the set A+B is a dodecad. From lemma (1.3) there are at most 132 different ways to generate the same dodecad. From lemma (1.2) there are 16 blocks intersecting a given block in a chosen pair of points. Consequently, there are at least $(759\cdot\binom{8}{2}\cdot 16)/132$ dodecads.

□

(1.7) Lemma: G_{24} is self-dual, i.e., $G_{24}=G_{24}^{\perp}$, and its weight distribution is

1 vector of weight o,
759 vectors of weight 8,
2576 vectors of weight 12,
759 vectors of weight 16, and
1 vector of weight 24.

Proof: From lemma (1.2) two blocks of $\mathcal{D}$ intersect in an even number of points. It follows that $G_{24} \subseteq G_{24}^{\perp}$. $1+759+2576+759+1 = 4096 = 2^{12}$. Consequently, $\dim G_{24} = \dim G_{24}^{\perp} = 12$, and "at least" in observation (1.6) has to be replaced by "exactly".

□

(1.8) Definition: A parity check on an (n,k)-code $\mathcal{C}$ is any vector $\underline{h} \in GF(2)^n$ such that $(\underline{h},\underline{c}) = o$ for all codewords $\underline{c} \in \mathcal{C}$.

2 Threshold Decoding Of The Extended Binary Golay Code

As G_{24} is self-dual any set of codevectors can be chosen for parity checks on the code itself. Decoding a received vector $\underline{u} = \underline{c} + \underline{e}$ with $\underline{c} \in G_{24}$ and $\mathrm{wgt}(\underline{e}) \leq 4$ is performed step-by-step. In order to decode a given coordinate u_i as parity checks take the 253 vectors of weight 8 associated with the 253 blocks passing through u_i.
If $\mathrm{wgt}(\underline{e}) = o$ none of the parity checks fails.
If $\mathrm{wgt}(\underline{e}) = 1$ and $\mathrm{supp}(\underline{e}) = \{i\}$ all the 253 parity checks fail, while for $\mathrm{wgt}(\underline{e}) = 1$ and $\mathrm{supp}(\underline{e}) = \{j\}$ $(i \neq j)$ 77 parity checks fail (see lemma (1.1)).
In the case $\mathrm{wgt}(\underline{e}) = 2$ assume first $\mathrm{supp}(\underline{e}) = \{i,j\}$. From lemma (1.1) we deduce that among the 253 parity checks that check u_i there are 77 that check u_j and 253-77=176 that don't check u_j. Hence, in this case, there fail 176 parity checks. Now assume $\mathrm{supp}(\underline{e}) = \{j,k\}$ $(k \neq i,j)$. From lemma (1.1) we deduce that among the 253 parity checks that check u_i there are 21 that check u_j and u_k, 77-21=56 that check u_j but not u_k, 56 that check u_k but not u_j, and 253-21-56-56=12o that check neither u_j nor u_k. Hence, there fail 56+56=112 parity checks in this case.
Treating the other cases similarly we obtain the following table:

	wgt($\underline{e}$)	$\#\{(\underline{1}_{B_i}, \underline{u}) = 1\}$
$e_i = o$	o	o
	1	77
	2	112
	3	125
	4	128
$e_i = 1$	4	128
	3	141
	2	176
	1	253

This table shows that decoding can be performed by a simple threshold test: set e_i=1 if at least 141 parity checks fail,
set e_i=o if at most 125 parity checks fail, and
4 errors have occured if exactly 128 parity checks fail.

The decoding algorithm is as follows:

```
FOR  i:=1  TO  24  DO
     z:=o
     FOR  B ∈ 𝔅_i  DO
          IF  wgt(u AND 1_B) odd  THEN
               z:=z+1
               IF  z > 128  THEN
                    u:=u+e_i              ; correcting coordinate u_i
                    EXIT
               FI
          FI
     OD
     IF  z = 128  THEN
          EXIT                            ; 4 errors detected
     FI
OD
```

3 Some Remarks On The Implementation

The threshold decoding algorithm has been implemented using a standard microprocessor unit (μP 8o85). Essential part of this machine is a table of the 759 blocks of the large Witt design arranged in a certain order, to the effect that the first 22 blocks restricted on the first 12 points yield a Hadamard design $S_2(3,6;12)$ (cf. [1]), and the first 132 blocks restricted on the first 12 points yield a Steiner system S(5,6;12) (cf. [1]).

The machine is able to

- display the blocks of the designs S(5,8;24), S(5,6;12), and $S_2(3,6;12)$,
- decode a given input $\underline{u}$ into a codeword $\underline{c} \in G_{24}$, a feature which obviously includes
    - finding the unique block of the Steiner system S(5,8;24) that contains five given points,
    - finding the unique block of the Steiner system S(5,6;12) that contains five given points out of a dodecad (for this purpose, two of the dodecads are marked with colors) (cf. [1]).

References

[1] T. Beth, D. Jungnickel
Mathieu groups, Witt designs, and Golay codes
(this volume)

[2] R.T. Curtis
A new combinatorial approach to $\mathcal{M}_{24}$
Mathematical Proceedings of the Cambridge Philosophical Society, 79 (1976), 25-42

[3] W. Fumy
Untersuchungen zum Codieren und Decodieren binärer Golay-Codes
Diplomarbeit, Universität Erlangen 198o

[4] I.B. Gibson, I.F. Blake
Decoding the binary Golay code with Miracle Octad Generators
IEEE Transactions on Information Theory, 24 (1978), 261-264

[5] J.M. Goethals
On the Golay perfect binary code
Journal of Combinatorial Theory, 11 (1971), 178-186

[6] N.J.A. Sloane
A short course on error-correcting codes
CISM Courses and Lectures No. 188
Springer, New York, 1975

[7] E. Witt
Die 5-fach transitiven Gruppen von Mathieu
Über Steinersche Systeme
Abhandlungen aus dem Mathematischen Seminar der Universität
Hamburg, 12 (1938), 256-264, 265-275

Walter Fumy
Institut für Mathematische Maschinen
und Datenverarbeitung I der
Universität Erlangen-Nürnberg
Martensstraße 3
D-852o Erlangen

k-DIFFERENCE-CYCLES AND THE CONSTRUCTION OF CYCLIC t-DESIGNS

Egmont Köhler
Math.Sem.Univ.Hamburg
Bundesstr.55
2000 Hamburg 13, Germany

Introduction

One of the most important methods to construct (v,k,λ)-blockdesigns (BIBD-s) comes from the application of the theory of difference families. However, the possibility of these constructions depends on the fact, that these designs are only twofold balanced. But in the general case of t-(v,k,λ)-designs ($S_\lambda(t,k,v)$), this method is not applicable. In this paper a necessary and sufficient condition for the existence of cyclic $S_\lambda(t,k,v)$ is proved (theorem 1), which yields an effective method to construct such systems. In some previons articles (e.g. [3],[4]) this method was considered mainly in the case that $v=p$ is a prime number, and e.g. numerous cyclic quadruple systems have been constructed in this way.
Here we deal with arbitrary parameters, and we show as an example, that the existence of cyclic 3-$(v,4,3)$-designs without repeated blocks for all $v\equiv 2 \pmod 4$, $v>2$ is a direct consequence of our theorem.

Some Definitions

Let $M = \{m_1, \ldots, m_n\} \neq \emptyset$ be a finite set and $\mathfrak{M}$ be a multiset over M, in which each $m_i \varepsilon M$ has multiplicity β_i.

Then we write

$$\mathfrak{M} = \sum_{i=1}^{m} \beta_i m_i .$$

Using this notation there is a natural way to define the sum of two multisets and the multiplication of a multiset whit a natural number. Furthermore we define

$$|\mathfrak{M}| := \sum_{i=1}^{m} \beta_i .$$

Now we consider natural numbers λ, t, k, v with $1<t<k<v$ and a set V with $|V| = v$.

Then let $\mathfrak{B} = \sum_{i=1}^{n} \beta_i B_i$ be a multiset over $\binom{V}{k}$, the set of all subsets of size k of V. If T is an arbitrary element of $\binom{V}{k}$, we define

$$\mathfrak{B}_T := \sum_{i=1}^{n} \beta_i^T B_i$$

by

$$\beta_i^T := \begin{cases} \beta_i, \text{if } T \subseteq B_i \\ o \quad \text{otherwise.} \end{cases}$$

Such a multiset $\mathfrak{B}$ is called a <u>t-design</u> (or more exactly: t-(v,k,λ)-design, in short $S_\lambda(t,k,v)$), iff for all $T \varepsilon \binom{V}{t}$ holds

$$|\mathfrak{B}_T| = \lambda .$$

Moreover, we call a $\mathfrak{B} = S_\lambda(t,k,v)$ <u>simple</u>, if $\mathfrak{B}$ is a set. The B_i' s, which are the terms of a sum in $\mathfrak{B}$, are the <u>blocks</u>, and the elements of V are the <u>points</u> of the $S_\lambda(t,k,v)$. A bijection $\mu: V \longrightarrow V$ is called an <u>automorphism</u> of $\mathfrak{B} = \sum_{i=1}^{n} \beta_i B_i = \sum_{i=1}^{n} \beta_i \{b_1^i, \ldots, b_k^i\}$ iff for every $i \varepsilon \{1, \ldots, n\}$:

1.) there exists a $j\varepsilon\{1, \dots ,n\}$, so that

$\{\mu(b_1^i), \dots ,\mu(b_k^i)\} = \{b_1^j, \dots ,b_k^j\}$, and

2.) $\beta_i = \beta_j$.

A $S_\lambda(t,k,v)$ is said to be cyclic, iff its automorphismgroup contains a cycle of length v.

k-Difference-cycles over $\mathbb{Z}_v$.

$\mathbb{Z}_v$ here denotes the additive group of the residues mod v, and the elements of $\mathbb{Z}_v$ are represented by the nonnegative numbers $0 \leq i < v$.

Clearly $\mathbb{Z}_v$ acts on $\binom{\mathbb{Z}_v}{k}$ by

$$x+B := x+\{b_1, \dots ,b_k\} := \{x+b_1, \dots ,x+b_k\}.$$

An orbit $B^+ \subseteq \binom{\mathbb{Z}_v}{k}$, which is generated by $B\varepsilon\binom{\mathbb{Z}_v}{k}$ in this way, can be described as follows: Let $B=\{b_1, \dots ,b_k\}$, $C=\{c_1, \dots ,c_k\}$ be elements of $\binom{\mathbb{Z}_v}{k}$ and $b_1<b_2< \dots <b_k$ and $c_1<c_2< \dots <c_k$. Then one has $C \in B^+$ iff the two vectors

$$(b_2-b_1,b_3-b_2, \dots ,b_k-b_1) \quad \text{and} \quad (c_2-c_1,c_3-c_2, \dots ,c_k-c_1)$$

differ only in a cyclic permutation of their components. Therefore, if one considers two given elements from the set

$$T_{k,v} := \{(a_1, \dots ,a_k)\varepsilon\,\mathbb{Z}_v^k \mid a_1+a_2+ \dots +a_k=v\}$$

to be equivalent ($a\sim b$) iff they differ only in a cyclic permutation of their components, then there exists a bijection

$$\phi_k : T_{k,v}/_\sim \longrightarrow \mathcal{L}_{k,v},$$

where $\mathcal{L}_{k,v} := \{B^+ \mid B\varepsilon\binom{\mathbb{Z}_v}{k}\}$.

This bijection evidently can be defined on the representatives:

$$\phi_k((a_1, \ldots, a_k)) := \{o, a_1, a_1+a_2, \ldots, \sum_{i=1}^{k-1} a_i\}.$$

Setting $T_{k,v}/\sim =: \tilde{T}_{k,v}$, we introduce

Definition 1 : An element $\kappa \in \tilde{T}_{k,v}$ is called a k-difference cycle over $\mathbb{Z}_v$.

If there is no danger of confusion, we will describe a cycle $\kappa \in \tilde{T}_{k,v}$ by a suitable generating element of $T_{k,v}$.

Now suppose v and k to have a nontrivial common divisor. Then there is some $B \in \binom{\mathbb{Z}_v}{k}$ for which $\ell_B := |B^+| < v$. In order to examine how such a behavior of B^+ can be rediscovered from $\phi_k^{-1}(B^+) \in \tilde{T}_{k,v}$, we consider

Definition 2: Let κ be a k-difference cycle over $\mathbb{Z}_v$. The smallest natural number p_κ having the property

$$\kappa = (a_1, \ldots, a_{p_\kappa}, a_1, \ldots, a_{p_\kappa}, \ldots, a_1, \ldots, a_{p_\kappa})$$

is called the period of κ.

Now it is easy to show:

Lemma 1: If $\kappa \in \tilde{T}_{k,v}$ and $\phi_k(\kappa) =: B^+$, then $\ell_B = p_\kappa \cdot \frac{v}{k}$.

Proof: Consider

$$\kappa = (a_1, \ldots, a_s, a_1, \ldots, a_s, \ldots, a_1, \ldots, a_s) \in \tilde{T}_{k,v},$$

$r := \frac{k}{s}$ and $\phi_k(\kappa) = B^+$, so that

$$B = \{o, a_1, a_1+a_2, \ldots, \sum_{i=1}^{k-1} a_i\}.$$

Since $\sum_{i=1}^{k} a_i = v$, it follows $\sum_{i=1}^{s} a_i = \frac{v}{r}$. Therefore

$$B = \{o, a_1, \ldots, \frac{v}{r} - a_s, \frac{v}{r}, \frac{v}{r} + a_1, \ldots, \frac{(r-1)v}{r}, \frac{(r-1)v}{r} + a_1, \ldots, v - a_s\},$$

and as an immediate consequence we have

$$x + B = B \quad \text{for} \quad x = v - \frac{(r-1)v}{r},$$

and x is the smallest natural number having this property. This yields $x = \frac{v}{r} = \frac{v \cdot s}{k} = p_{\kappa} \cdot \frac{v}{k}$.

Looking at $B \in \binom{\mathbb{Z}_v}{k}$, every element from B^+ contains exactly $\binom{k}{t}$ pair-wise distinct t-elementary subsets. Therefore the totality of these $\ell_B \cdot \binom{k}{t}$ t-subsets of elements of B^+ form a multiset, which we will denote by B^t.

Now it is clear, that, if $T \in \binom{\mathbb{Z}_v}{k}$ has multiplicity r in B^t, so also has any other element $\in T^+$. This fact shall be denoted here by

$$r.T \in B^t \Longleftrightarrow r.T^+ = B^t.$$

What is now the meaning of $r.T^+ \subseteq B^t$ in terms of $\phi_t^{-1}(T^+)$ and $\phi_k^{-1}(B^+)$?

To answer this question, first of all we give the following

<u>Definition 3:</u> Let $\kappa = (a_o, \ldots, a_{k-1}) \in \tilde{T}_{k,v}$ and $\tau = (b_o, \ldots, b_{t-1}) \in \tilde{T}_{t,v}$. Then τ is a <u>t-subcycle of κ</u> (for short: $\tau < \kappa$), iff both 1.) and 2.) hold :

1.) $\bigwedge_{i \in \mathbb{Z}_t} \bigvee_{j,j^* \in \mathbb{Z}_k} \quad b_i = \sum_{h=j}^{j^*} a_h$

2.) $\left(b_i = \sum_{h=j}^{j^*} a_k \wedge b_{i+1} = \sum_{h=p}^{p^*} a_k \right) \Longrightarrow \left(p = j^* + 1. \right)$

Using this definition, one can show:

<u>Lemma 2:</u> If $B \in \binom{\mathbb{Z}_v}{k}$ and $T \in \binom{\mathbb{Z}_v}{k}$, then

$$T^+ \subseteq B^t \Longleftrightarrow \phi_t^{-1}(T) < \phi_k^{-1}(B) .$$

<u>Proof:</u> Let $B = \{b_1, \ldots, b_k\}$ be given and $b_1 < \ldots < b_k$. Then $T \subseteq B$ if $T = \{b_{i_1}, \ldots, b_{i_t}\}$ and $i_1 < \ldots < i_t$.

Now by definition 1:

$\phi_k^{-1}(B) = \Big(b_{i_1+1} - b_{i_1}, b_{i_1+2} - b_{i_1+1}, \ldots, b_{i_2} - b_{i_2-1}, b_{i_2+1} - b_{i_2}, b_{i_t} - b_{i_t-1},$

$$b_{i_t+1}-b_{i_t}, \ldots, b_1-b_k, b_2-b_1, \ldots, b_{i_1}-b_{i_1-1}\Big\},$$

and, considering the indices as elements of $\mathbb{Z}_k$, we have

$$\left\{\sum_{j=1}^{i_2-i_1} b_{i_1+j}-b_{i_1+j-1}, \sum_{j=1}^{i_3-i_2} b_{i_2+j}-b_{i_2+j-1}, \ldots, \sum_{j=1}^{i_1-i_t} b_{i_t+j}-b_{i_t+j-1}\right\}$$

$$=\left\{b_{i_2}-b_{i_1}, b_{i_3}-b_{i_2}, \ldots, b_{i_1}-b_{i_t}\right\} = \phi_t^{-1}(T).$$

Conversely, if

$$\tau = \left(\sum_{i=1}^{j_1} a_i, \sum_{i=j_1+1}^{j_2} a_i, \ldots\right) < \kappa = (a_1, \ldots, a_k),$$

then $\phi_k(\kappa) = \{o, a_1, a_1+a_2, \ldots, \sum_{i=1}^{k-1} a_i\}$, and therefore $\phi_t(\tau) \subseteq \phi_k(\kappa)$ is immediate.

With regard to definition 3, each $\kappa \in \tilde{T}_{k,v}$ contains exactly $\binom{k}{t}$ (not necessarily pairwise distinct) t-subcycles. In the following we will denote by ${}^t\kappa$ the multiset of all t-subcycles of κ. Furthermore, we denote by κ_τ the multiplicity of $\tau \in {}^t\kappa$.

Now we can prove:

<u>Lemma 3:</u> Let $B \in \binom{\mathbb{Z}_v}{k}$, $T \in \binom{\mathbb{Z}_v}{t}$, $\kappa := \phi_k^{-1}(B^+)$ and $\tau = \phi_t^{-1}(T^+)$. Then, if r is maximal with $r \cdot T^+ \subseteq B^t$, the following equation holds:

$$r = \frac{t}{k} \cdot \frac{p_\kappa}{p_\tau} \cdot \kappa_\tau.$$

<u>Proof:</u> For T, B and $r \cdot T^+ \subseteq B^t$ we have $r \cdot T \in B^t$. Now let $B = \{a_1, \ldots, a_k\}$ and $T = \{a_{i_1}, \ldots, a_{i_t}\}$. First of all, we consider the case, that there is no $c \in \mathbb{Z}_v \setminus \{o\}$, with $c+T = B$.

Since $|\{c \in \mathbb{Z}_v \mid c+T = T\}| = \frac{v}{\ell_T}$, we have $r = \frac{\ell_B}{\ell_T}$.

In the other case, there are $c_1, \ldots, c_s \in \mathbb{Z}_v - \{o\}$ and $c_i + T \subseteq B$ for $i \in \{1, \ldots, s\}$. Then $r = \frac{\ell_B}{\ell_T} \cdot s$. From definition 3 and the proof of Lemma 2 we get $s = \kappa_\tau$.

So by Lemma 1 we conclude from $r = \frac{\ell_B}{\ell_T} \cdot \kappa_\tau$ the assertion.

Lemma 3 now suggests the following

Definition 4: For $\kappa \in \tilde{T}_{k,v}$ we define the following multiset:

$$\kappa^t := \frac{t}{k} \cdot p_\kappa \cdot \sum_{\tau < \kappa} \frac{\kappa_\tau}{p_\tau} \cdot \tau .$$

Thinking back to the definition of a cyclic $S_\lambda(t,k,v)$, we now have proved the announced

Theorem 1: A cyclic $S_\lambda(t,k,v)$ exists iff there is a multiset $\sum_{i=1}^{n} \beta_i \kappa_i$ over $\tilde{T}_{k,v}$, so that

$$\sum_{i=1}^{n} \beta_i \kappa_i^t = \lambda \cdot \tilde{T}_{t,v} .$$

Symmetric Difference cycles and

Quadruple systems.

A k difference cycle κ is called <u>symmetric,</u> iff
$\kappa = (a_1, a_2, \ldots, a_k) = (a_1, a_k, a_{k-1}, \ldots, a_2)$.

By $s\tilde{T}_{k,v}$ we denote the set of all symmetric k-difference cycles over $\mathbb{Z}_v$. Since every $\kappa \in s\tilde{T}_{4,v}$ is of the form $\kappa = (a,b,a,c)$, it is easily shown, that $|s\tilde{T}_{k,v}| = \binom{\frac{v}{2}}{2}$ if $v \equiv 2 \pmod 4$, and in the remainder of this section we consider this situation.

Then $s\tilde{T}_{4,v}$ contains exactly $\frac{v-2}{4}$ elements κ, having the property $p_\kappa \neq 4$ and $p_\kappa = 2$, namely $\kappa = (a,b,a,b)$ and $a+b=\frac{v}{2}$. As a consequence of

Lemma 1, for these κ-s we have $\ell_B = \frac{v}{2}$, if we denote $\phi^{-1}(\kappa) =: B^+$.

It follows:

$$\sum_{\kappa \in \tilde{T}_{4,v}} |\phi^{-1}(\kappa)| = v \cdot \left(\binom{\frac{v}{2}}{2} - \frac{v-2}{4} \right) + \frac{v}{2} \cdot \frac{v-2}{4} = 3 \cdot \frac{v(v-1)(v-2)}{4 \cdot 3 \cdot 2},$$

and the latter is just the number of blocks of an $S_3(3,4,v)$.

That means: if we can show, that every element of $\tilde{T}_{3,v}$ is contained at least three times as a term of a sum in $\sum_{\kappa \in s\tilde{T}_{4,v}} \kappa^t$, then by theorem 1 the blocks of the orbits, described by the elements of $s\tilde{T}_{4,v}$, form a simple cyclic $S_3(3,4,v)$.

In order to prove this, we subdivide the set of 3-cycles into four classes:

1.) $\tau = (a,b,c)$ and $a<b<c$.
   Then $\tau \in \kappa^3$ for $\kappa = (a,b-a,a,c),(a,b,a,c-a),(a,b,c-b,b)$.
2.) $\tau = (a,a,b)$ and $a<b$.
   Then $\tau \in (a,a,\frac{b}{2},\frac{b}{2})^3$ and $2 \cdot \tau \in (a,a,a,b-a)^3$.
3.) $\tau = (a,a,b)$ and $a>b$.
   Then $\tau \in (a,a,\frac{b}{2},\frac{b}{2})^3$ and $2 \cdot \tau \in (b,a-b,b,a)^3$.
4.) $\tau = (a,a,a)$.
   Then $3 \cdot \tau \in (a,a,b,b)^3$, since $P_{(a,a,b,b)} = 2$ and $P_\tau = 1$.

As a result, we have

<u>Theorem 2:</u> If $v \equiv 2 \pmod 4$ and $v>2$, then there exists a simple cyclic $S_3(3,4,v)$.

Similarly, one can construct cyclic $S_3(3,4,v)$-s for all $v \equiv 0 \pmod 4$, but these designs are not simple. (See e.g. [2].)
Until now, it remains an open question, whether theorem 2 also holds for $v \equiv 0 \pmod 4$.

For example: $\sum_{i=1}^{6} \kappa_i^3$, defined by

$\kappa_1 = (1,1,2,4)$, $\kappa_2 = (1,1,3,3)$, $\kappa_3 = (1,1,4,2)$, $\kappa_4 = (1,2,1,4)$,

$\kappa_5 = (1,2,3,2)$ and $\kappa_6 = (2,2,2,2)$

form a simple cyclic $S_3(3,4,8)$ by theorem 1.

References

[1] J.DOYEN and A.ROSA:
An updated bibliography and survey of Steiner systems, Annals of Discr. Math. 7 (1980), 317-349

[2] M.KLEEMANN:
k-Differenzenkreise und 2-fach ausgewogene Pläne, Diplomarbeit, Universität Hamburg (1980)

[3] E.KÖHLER:
Zyklische Quadrupelsysteme, Abh. Math. Sem. Univ. Hamburg 48 (1979), 1-24

[4] E.KÖHLER:
2-auflösbare Quadrupelsysteme über $\mathbb{Z}_p$ submitted to J. Comb. Th.

[5] C.LINDNER and A.ROSA:
Steiner quadruple systems - a survey
Discrete Math. 22 (1978), 147-181.

CHARACTERIZATION OF BIPLANES

BY THEIR AUTOMORPHISM GROUPS

E.S. Lander
Wolfson College
and
The Mathematical Institute

Oxford, England

A central theme in the study of finite geometries has been the classification of projective planes according to their automorphism groups. We can ask similar questions more generally about symmetric (v,k,λ) designs with $\lambda>1$--however, rather different methods are usually required, and on the whole less is known. In this paper we explore which biplanes D (that is, symmetric designs with $\lambda=2$) possess the following properties:

PROPERTY A. *There exists an automorphism group* G *of* D *which fixes a block* B *and acts on the points of* B *as* PGL(2,q) *acts on* PG(1,q).

PROPERTY B. *There exists an automorphism group* G *of* D *which fixes a block* B *and acts on the points of* B *as* PSL(2,q) *acts on* PG(1,q).

Such biplanes would necessarily have the parameters $(v,k,\lambda) = (\frac{1}{2}(q^2+q+2),q+1,2)$.

The problem is interesting for a number of reasons. First of all, five of the sixteen known biplanes enjoy Properties A and B. They are:

(1) the unique biplane with $k=3$.

(2) the unique biplane with $k=4$.

(3) the unique biplane with $k=5$.

(4) the "nicest" biplane with $k=6$ (namely the one with a 2-transitive automorphism group).

(5) one of the four biplanes with $k=9$.

We might remark that in the first four cases, the full automorphism group is even larger than PGL(2,q), being in fact transitive on blocks. In the fifth example, PGL(2,8)=PSL(2,8) is the full automorphism group.

Full details of these biplanes can be found in [1].

The action of PSL(2,q) on these biplanes is more than incidental. Hall showed how each of the biplanes could be easily constructed by starting with the assumption that PSL(2,q) acts as a block stabilizer in a particularly nice fashion. He had hoped at first to produce an infinite series of biplanes with Property B in this manner, but instead he proved [5] that his particular construction failed for $k>9$. Hall suggested that perhaps no further biplanes with Property B exist.

By now, the question has become something of a loose end, since the analogous problem for $\lambda\neq 2$ has been settled.

THEOREM 1. Suppose that D is a symmetric (v,k,λ) design with Property B.

(1) If $\lambda=1$, then D is a Desarguesian projective plane.

(2) If $\lambda\geq 3$, then D is one of the following:

(a) the unique symmetric (11,6,3) design.

(b) the complementary design to PG(3,2).

(c) the complementary design to the "nicest" biplane with $k=6$.

Proof. (1) follows from Lüneburg [11] and Yaqub [15].

(2) has been studied extensively by Kantor [7]. Kelly [8] provided the complete answer.

In this paper we shall prove the following results.

THEOREM 2. A biplane with Property A has $k=2,3,4,5$, or 8.

THEOREM 3. A biplane with Property B and a transitive automorphism group has $k=2,3,4$, or 5.

THEOREM 4. Suppose that D is a biplane with Property B. Then

(1) if q is a power of 2, then $q=2,4$, or 8.

(2) if $q\equiv 3 \pmod 4$, then $q=3$.

(3) if $q\equiv 1 \pmod 4$, then $q=5$ or else q is a prime and PSL(2,q) acts as the full automorphism group of D.

Theorem 2 is in a very real sense due to Hall [5]. All we must do is to show that any biplane with Property A must be of the particular sort that Hall excluded. The most novel of the proofs we give is that of Theorem 4(3). It relies on techniques inspired by coding theory and involving modular representation theory. (The methods were first presented in [9] but this is their first application to a nonabelian group.)

1. Facts about symmetric designs. We assume that the reader is acquainted with the basic facts about symmetric designs. In particular, he should be familiar with the terms order, dual, complement, incidence matrix, automorphism group and full automorphism group of a symmetric design, as well as with the Bruck-Ryser-Chowla Theorem. As a reference, see [3,9,14].

2. Property A. The structure of a biplane can be described quite succinctly using only a single block B. (See, e.g. [1].) It is easy to represent the blocks distinct from B: any unordered pair of points shall represent the unique block which meets B in precisely these two points. How can we represent points off B? Let r be such a point. For any point p on B there are two blocks incident with r and p. Each meets B in one further point and the two points arising in this way are distinct. Define a graph $\Gamma(r)$ whose vertices are the points of B and in which $\{p_1,p_2\}$ is an edge if and only if there is a block incident with p_1,p_2 and r. The graph has valency 2 and so is a disjoint union of polygons. We represent the point r by this graph $\Gamma(r)$. The definition of a biplane implies that the collection Γ of graphs $\Gamma(r)$ with r off B has the following properties:

(1) for any three distinct points p_1,p_2,p_3 on B, $\{p_1,p_2\}$ and $\{p_1,p_3\}$ are both edges in a unique graph in Γ.

(2) for any four distinct points p_1,p_2,p_3,p_4 on B, $\{p_1,p_2\}$

and $\{p_3,p_4\}$ are both edges in just two graphs in Γ.

(3) Any two graphs in Γ have just two edges in common.

Conversely, given any collection of graphs of valency 2 on a k-set satisfying these statements we can recover a biplane. The representation is particularly convenient for studying automorphism groups G fixing a block B, since it relates the action of G on points and blocks not incident with B to the action of G on the points incident with B.

Let D be a biplane with Property A. Let G be the required group and, for simplicity, let us identify the points of B with the elements of PG(1,q). Since PGL(2,q) is transitive on unordered pairs from PG(1,q) the group G transitively permutes the blocks distinct from B. Now, an automorphism group of a symmetric design has an equal number of orbits on points and blocks. As G has two block orbits, it necessarily has two point orbits--which must be the points on B and off B. Since the points off B correspond to graphs in Γ, the action of G on the points of B must transitively permute the $\frac{1}{2}(q^2-q)$ graphs in Γ. Hence, the subgroup H of G stabilizing any particular graph $\Gamma(r)$ is a subgroup of PGL(2,q) having order 2(q+1) and index $\frac{1}{2}(q^2-q)$. From Dickson's list [4] of all subgroups of PGL(2,q) we can determine H. From H we can deduce the structure of $\Gamma(r)$.

<u>Case 1, when $q=2^e$.</u> According to Dickson's list, H must be a dihedral group of order 2(q+1) whose cyclic subgroup C acts transitively on the (q+1) points of PG(1,q). What graphs of valency 2 are fixed by such a cyclic group, or dihedral group? A picture makes the answer patent:

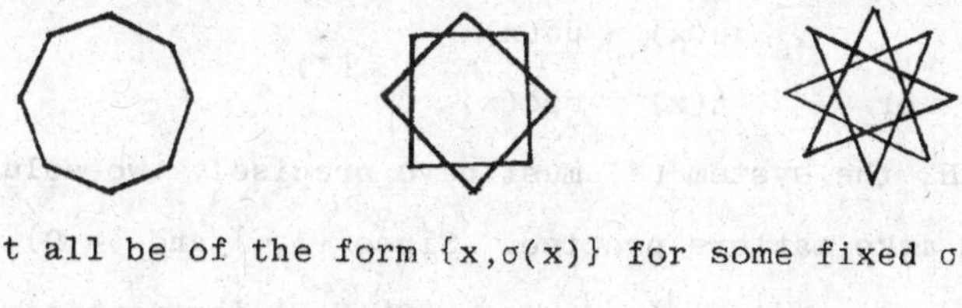

The edges must all be of the form $\{x,\sigma(x)\}$ for some fixed $\sigma \varepsilon C \subset PGL(2,q) = PSL(2,q)$.

<u>Case 2, when q is odd.</u> Except when q=11, 29, or 59, the group H

must be $Z_2 \cdot D_{q+1}$, where D_{q+1} is a dihedral group of order q+1 acting on two $\frac{1}{2}(q+1)$-gons (so to speak) and the Z_2 exchanges these two $\frac{1}{2}(q+1)$-gons. The reader should convince himself that except in the degenerate case when q=3 (which we henceforth ignore, since all biplanes with k=4 are known) $\Gamma(r)$ must have the following form: there are no edges between the two $\frac{1}{2}(q+1)$-gons and, there is a fixed integer i such that within each of the $\frac{1}{2}(q+1)$-gons all pairs of vertices i apart are joined. Thus, all edges have the simple form $\{x,\sigma(x)\}$ for some σ in the cyclic subgroup of D_{q+1}. In particular $\sigma \varepsilon PSL(2,q)$.

If q=11,29, or 59, there is the additional possibility that **H** is A_4, S_4, or A_5, respectively. However, since by the Bruck-Ryser-Chowla Theorem no biplanes exist with k=12,30 or 60, we need not worry about these possibilities.

In both Cases 1 and 2, the edges of $\Gamma(r)$ are of the simple form $\{x,\sigma(x)\}$ for some $\sigma\varepsilon PSL(2,q)$ (except when q=3). This is precisely the sort of biplane which Hall studied and completely classified in [5]. However, since Hall's original proof is quite long and fussy, I offer a much shorter proof, by invoking some algebraic geometry.

The edges of $\Gamma(r)$ have the form $\{x,\sigma(x)\}$ for a fixed $\sigma\varepsilon PSL(2,q)$. The other graphs in Γ are obtained as the image of $\Gamma(r)$ under the action on PGL(2,q) on the vertices. For each $\mu \varepsilon PGL(2,q)$ we have the graph $\Gamma^{\mu}(r)$ in which the edges are $\{\mu(x),\mu\sigma(x)\}$ for all $x\varepsilon PG(1,q)$. (Of course, elements μ and μ' in the same coset of H give the same graph.) Now, any two distinct graphs must share exactly two edges. A typical edge $\{\mu(x),\mu\sigma(x)\}$ of $\Gamma^{\mu}(r)$ is also an edge of $\Gamma(r)$ if either

$$\begin{aligned} \sigma\mu(x) &= \mu\sigma(x) \\ \text{or} \quad \mu(x) &= \sigma\mu\sigma(x) \end{aligned} \qquad (*)$$

Thus for $\mu\notin H$, the system (*) must have precisely two solutions in PG(1,q

Let us make matters precise. Since $\{1,\infty\}$ and $\{\infty,0\}$ occur as edges in some graph in Γ (and since PGL(2,q) is transitive on Γ), there is no harm in supposing that they are edges in $\Gamma(r)$ and that $\sigma(1)=\infty$ and

$\sigma(\infty)=0$. Then σ has the form

$$\sigma: x \to \frac{a}{1-x}$$

for some $a \in F_q$. Let μ be the transformation $x \to ux$ for $u \in F_q$ and $u \neq 0,1,-1$ Clearly, $\mu \notin H$. The first equation has two solutions, $x=\infty$ and $x= 1/(u+1)$. Therefore the second equation must have no solutions. Now, any solution of the second equation satisfies

$$0 = ux^2+(u^2a-a-u)x+a.$$

If we write $f(u,x) = ux^2+(u^2a-a-u)+a$ then we have determined that $f(u,x)=0$ must have no solutions except if $u=0,1,-1$. For $u=0$, it clearly has one solution; for $u=1$ or -1 it has at most two solutions Thus, in all $f(u,x)=0$ has at most five solutions in $F_q xF_q$. This is a lot to ask! The equation $f(u,x)=0$ describes an elliptic curve in the two-dimensional plane over F_q. Algebraic geometers have excellent estimates for the number N of points on such a surface. Weil showed (see [6]) that

$$|N-q| \le 2\sqrt{q}.$$

Now, we have $|5-q| > 2\sqrt{q}$, except for $q \le 11$. Thus, a necessary condition for the existence of a biplane with Property A is that $q \le 11$. The possibilities are $q=2,3,4,5,7,9,11$. Since all biplanes with $k=3,4,5,6,8,10,12$ are known, it is just a matter of checking the list to prove Theorem 2.

<u>3. Property B, the case $q=2^e$.</u> When q is a power of 2, we have PGL(2,q)=PSL(2,q) and there is nothing more to prove. When q is odd, however, PSL(2,q) is a subgroup of index 2 in PGL(2,q). Mimicing the proof of the previous section, we find that H is a dihedral group of order (q+1). (If $q \equiv 1 \pmod 4$, then H acts intransitively as a dihedral group on two $\frac{1}{2}(q+1)$-gons. If $q \equiv 3 \pmod 4$, then the cyclic subgroup of index 2 in H acts regularly on the two $\frac{1}{2}(q+1)$-gons and any further element exchanges these orbits.) While it is not hard to determine the possible graphs $\Gamma(r)$ the edges no longer need be

of the simple form $\{x,\sigma(x)\}$ for a particular $\sigma \varepsilon PSL(2,q)$. Accordingly, we need a different approach.

4. Property B, the case $q \equi 3 \pmod 4$. Although the incidence matrix of a symmetric (v,k,λ) design F is quite a useful tool, a slightly different matrix suits me better. I call the $(v+1)x(v+1)$ matrix

$$B = \left(\begin{array}{c|c} A & \begin{matrix} 1 \\ \vdots \\ 1 \end{matrix} \\ \hline \lambda \dots \lambda & k \end{array} \right)$$

the extended incidence matrix of F. Also, let ψ be the diagonal matrix $Diag(1,1,\dots,1,-\lambda)$ of size $(v+1)$. From the equation $AA^T = nI+\lambda J$ for the incidence matrix, we have

$$B\psi B^T = n\psi .$$

The extended Z-module of F is defined to be the Z-module M obtained by taking all integral linear combinations of the rows of B. We regard M as a submodule of $W=Z^{v+1}$, the Z-module of all integral $(v+1)$-tuples.

The module M can be endowed with a great deal of structure inherited from F. If G is any automorphism group of F, then permuting the columns (i.e., points) of the incidence matrix A according to G has the same effect as rearranging the rows (i.e., blocks). The same is true of the extended incidence matrix B once we adopt the convention that G fixes the final column. Since these rows generate M, the module M is closed under the permutation action of G on its coordinates. This G-action clearly fixes $W=Z^{v+1}$ as well. So, we can regard M as a sub-ZG-module of W. (Note: all modules will be left modules.)

Often it is easier to work over a finite field. For a prime p, let $\pi_p : Z^{v+1} \to F_p^{\ v+1}$ be the "reduction mod p" homomorphism. Then $\pi_p(M)$ is a sub-F_pG-module of $\pi_p(W)$, with G having the same action on coordinates.

In the problem at hand, our plan will of course be to obtain PSL(2,q)-modules and to study them. Suppose now that D is a biplane with Property B and that $q\equiv 3 \pmod 4$. Rather than working with the extended module of D we prefer to set $F = (D^{dual})^{complement}$ and work with the extended Z-module M of F. For reference, F has parameters $(\frac{1}{2}(q^2+q+2), \frac{1}{2}(q^2-q), \frac{1}{2}(q-1)(q-2)) = (v, k', \lambda')$.

The reason for using D^{dual} is that G acts on the coordinates of the extended module of a design as it acts on the points of the design. In D, however, the block-action of PSL(2,q) is easier to fully describe: the group G=PSL(2,q) acts by fixing a block B and permuting the other blocks as PSL(2,q) permutes unordered pairs from PG(1,q). This block-action becomes the point-action in the dual design.

We further take the complement, $(D^{dual})^{complement}$, in order to make the following proposition true.

PROPOSITION 6. <u>With respect to the ordinary dot product,</u> $\pi_2(M)$ <u>is a totally isotropic subspace of</u> $\pi_2(W)$ <u>having dimension</u> $\frac{1}{2}(v+1)$. <u>In other words,</u> $\pi_2(M)^{\perp} = \pi_2(M)$ <u>(where "$\perp$" denotes the orthogonal subspace with respect to the ordinary dot product).</u>

<u>Proof.</u> Reduce modulo 2 the equation

$$B\psi B^T = n\psi .$$

Since λ' is odd, $\psi \equiv I \pmod 2$. Moreover, since $n=q-1$ is even, $n\equiv 0 \pmod 2$. Hence

$$BB^T \equiv 0 \pmod 2.$$

This says that the dot product of rows of B is even. Hence $\pi_2(M)$ is totally isotropic.

Since $\pi_2(M)$ is totally isotropic with respect to a nonsingular scalar product, its dimension is at most $\frac{1}{2}(v+1)$. To show that it is at least $\frac{1}{2}(v+1)$, we need to use the fact that $2||n$ (which is only true when $q\equiv 3 \pmod 4$). From the equation $B\psi B^T=n\psi$, we see that $|\det B| = n^{\frac{1}{2}(v+1)}$. So, $2^{\frac{1}{2}(v+1)}||\det B$. Now from the theory of invariant factors, there exist unimodular integral matrices P and Q such that

$$PBQ = \begin{pmatrix} d_1 & & \\ & \ddots & \\ & & d_{v+1} \end{pmatrix} := E$$

The dimension of $\pi_2(M)$ is the F_2-rank of B, which is also the F_2-rank of E. Since $2^{\frac{1}{2}(v+1)} \| \det B = \det E$, at most $\frac{1}{2}(v+1)$ of the diagonal entries d_i are even. Hence, the dimension of $\pi_2(M)$ is at least $\frac{1}{2}(v+1)$, proving the proposition. □

Now, consider the situation. We have a vector space $V=\pi_2(W)$ equipped with the dot product. The group $G=PSL(2,q)$ acts on V and preserves the scalar product (every permutation of the coordinates does). Moreover, we have a subspace $C=\pi_2(M)$ which is closed under the G-action and which satisfies $C^{\perp} = C$. Let us begin to make a composition series for V as an F_2G-module. We start with

$$0 \subset C \subset V.$$

If we fill in a composition series for C, we can obtain a composition series for the portion between C and V simply by taking the orthogonal subspaces of those in the first half. To wit, we have a composition series of the form

$$0 = C_0 \subset C_1 \subset \dots \subset C_t=C=C^{\perp}=C_t^{\perp} \subset \dots \subset C_1^{\perp} \subset C_0^{\perp} = V.$$

A brief word about representation theory is now in order. Let F be a field and let H be a group. If T is a left FH-module, we define T* to be the span of the F-linear functionals on T. That is, $T^*=\mathrm{Hom}_F(T,F)$. We can give T* a left FH-module structure by the rule

$$(h\eta)(v) = \eta(h^{-1}v)$$

for $h\in H$, $\eta\in T$, and $v\in T$. We call T* the <u>contragredient</u> of T. It is also easy to describe contragredients in terms of matrix representations. If ρ is a matrix representation of the action of G on T then ρ^* is given by $\rho^*(g)= \rho(g^{-1})^T$. Thus, if χ is the character of a representation, the character χ^* of its contragredient is given by $\chi^*(g)=\chi(g^{-1})$. An irreducible representation is called <u>self-contragredie</u> if it is isomorphic to its contragredient; an irreducible representation with character χ is self-contragredient if and only if $\chi(g)= \chi(g^{-1})$ for all $g\in H$.

Now, let us return to our composition series above. By formal algebra we can show (see [9]) that

$$(C_i/C_{i-1})^* \simeq C_{i-1}^{\perp}/C_i^{\perp}.$$

Hence everytime an irreducible module occurs as a composition factor in the first half, its contragredient occurs in the second half. The conclusion we draw is that any self-contragredient module must occur with even multiplicity as a composition factor.

All that remains is for us to determine the composition factors of our module $V=\pi_2(W)$. Actually, it is simpler for us to work over an algebraically closed field, so we shall determine the composition factors of $\Omega_2 \otimes \pi_2(W)$, where Ω_2 is an algebraic closure of F_2. (Self-contragredient composition factors of $\Omega_2 \otimes \pi_2(W)$ must also occur with even multiplicity of course.) We now do so.

We start with the representation theory of PSL(2,q) over the complex numbers and pass from ordinary characters to modular characters. The conjugacy classes of PSL(2,q) are well known. Let

<a> be a cyclic subgroup of PSL(2,q) of order $\frac{1}{2}(q-1)$;

<b> be a cyclic subgroup of PSL(2,q) of order $\frac{1}{2}(q+1)$;

q_1 and q_2 be representatives of the two classes of elements with order dividing q.

Then every element of PSL(2,q) is conjugate to one of 1, q_1, q_2, a^i (with $1 \le i \le \frac{1}{2}(q-1)$), or b^j (with $1 \le j \le \frac{1}{2}(q+1)$). Of these, a^i is conjugate only to a^{-i} and b^j only to b^{-j}.

A complete ordinary character table for PSL(2,q) with $q\equiv 3 \pmod 4$ is given in Table 1.

Since W is a permutation module its character χ is easy to find. For $g \in G$, the value $\chi(g)$ is the number of coordinates fixed by g. We calculate that

$\chi(1)= \frac{1}{2}(q^2+q+4)$; $\quad\quad$ $\chi(a^i)= 3$ for $i=1,\dots,\frac{1}{2}(q-3)$

$\chi(q_1)= 2$; $\quad\quad$ $\chi(b^j)= 2$ for $j=1,\dots,\frac{1}{2}(q-3)$

$\chi(q_2)= 2$; $\quad\quad$ $\chi(b^{\frac{1}{2}(q+1)})= \frac{1}{2}(q+5)$.

	1	q_1	q_2	a^i, $i=1,\ldots,\frac{1}{4}(q-3)$	b^j, $j=1,\ldots,\frac{1}{4}(q+1)$
ζ_0	1	1	1	1	1
ζ_1	q	0	0	1	-1
ζ_2	$\frac{1}{2}(q-1)$	$\frac{-1+\sqrt{-q}}{2}$	$\frac{-1-\sqrt{-q}}{2}$	0	$(-1)^{j+1}$
ζ_3	$\frac{1}{2}(q-1)$	$\frac{-1-\sqrt{-q}}{2}$	$\frac{-1+\sqrt{-q}}{2}$	0	$(-1)^{j+1}$
Ψ_R, $1\leq R\leq\frac{1}{4}(q-3)$	$q+1$	1	1	$\omega^{iR}+\omega^{-iR}$	0
Φ_S, $1\leq S\leq\frac{1}{4}(q-3)$	$q-1$	-1	-1	0	$-(\theta^{js}+\theta^{-js})$

where ω is a primitive $\frac{1}{2}(q-1)$-st root of unity and θ is a primitive $\frac{1}{2}(q+1)$-st root of unity.

Table 1.

The inner product

$$\langle\chi,\eta\rangle = \frac{1}{|G|}\sum_{g\varepsilon G}\chi(g)\bar{\eta}(g)$$

counts how often the irreducible character η figures in χ. We calculate directly that Ψ_1 figures exactly once in χ. (Note that Ψ_1 only appears as an irreducible character in the character table when $q>3$. Hence we must exclude the case $q=3$ in what follows.) Moreover Ψ_1 is a self-contragredient character.

Let $\hat{\Psi}_1$ be the reduction of Ψ_1 modulo 2. (For a reference on modular representation theory, see [13].) We must first show that $\hat{\Psi}_1$ is an irreducible modular character and then determine how often it figures as a composition factor of $\Omega_2\otimes\pi_2(W)$. The next proposition supplies the answer.

PROPOSITION 7. Let ζ be an absolutely irreducible ordinary character of a group G and let r be a prime. Suppose that the highest power of r dividing $|G|$ and the highest power of r dividing $\zeta(1)$ are the same. Then the reduction $\hat{\zeta}$ of ζ mod r is an absolutely irreducible character. Moreover, if η is any other absolutely irreducible ordinary character, then $\hat{\eta}$ does not involve $\hat{\zeta}$.

Proof. See [13].

In our case, the same power of 2 exactly divides $\Psi_1(1)$ and $|PSL(2,q)|$. Thus, $\hat{\Psi}_1$ is an irreducible modular character. By the "moreover" part of the proposition, $\hat{\Psi}_1$ figures as a character of $\Omega_2 \otimes \pi_2(W)$ exactly as often as Ψ_1 figures as a character of W; that is, it figures with multiplicity one.

However, this is a contradiction. For, $\hat{\Psi}_1$ is a self-contragredient character and, by the discussion above, it must figure with even multiplicity. We have reached a contradiction by assuming that there exists a biplane with Property B where $q \equiv 3 \pmod 4$ and $q \neq 3$. Hence, a biplane with Property B and $q \equiv 3 \pmod 4$ must have $q=3$.

5. Property B, the case $q \equiv 1 \pmod 4$. It would be nice to apply a similar argument in the case $q \equiv 1 \pmod 4$. However, we are not so fortunate. On the face of things, the crucial Proposition 6 seems to require the stringent condition that $2||n$. Lander [9] has generalized this, showing how to perform a similar construction for any prime dividing the square-free part of n. Unfortunately, even this is not enough. For, when $q \equiv 1 \pmod 4$, the number of points in the design is even and hence by the first part of the Bruck-Ryser-Chowla Theorem (also known as Schutzenberger's Theorem) the order n must be a square. Disappointing as this is, we can turn it to our advantage to obtain a partial resolution of this case.

Consider a biplane satisfying Property B with $q \equiv 1 \pmod 4$. Since all biplanes with k=6 are known, we assume that q>5. The order n equals q-1. Since it is a square, $q=4x^2+1$. Lesbesgue showed in 1850 that a prime power of the form $4x^2+1$ must be simply a prime [10]. Say q=p.

Now, let K be the subgroup of the full automorphism group stabilizing the block B (this might be larger than G = PSL(2,p)). The group K acts faithfully on the points of B. Considering K as a permutation group on the p+1 points of B, set $K' = A_{p+1} \cap K$. We have $PSL(2,p) \subseteq K' \subseteq A_{p+1}$. Now, Neumann [12] has shown that if X is any group such that $PSL(2,p) \subsetneqq X \subseteq A_{p+1}$ and p>7, then X must be 4-transitive.

Thus, either K'=PSL(2,p) or K' acts 4-transitively on B. By using the determination of all perfect error-correcting codes, Cameron [2] has determined all biplanes admitting a group fixing a block and acting 4-transitively on the points of the block. Such a biplane has at most 6 points on a block. Hence, in our case we may conclude that K' = PSL(2,p). (Actually, it is not necessary to invoke Cameron's nice result. For, the subgroup of K' stabilizing three points fixes one of the graphs $\Gamma(r)$ and is still transitive on the remaining points. Hence, $\Gamma(r)$ must be either a single quadrilateral or else a union of triangles. By Section 3, the graph $\Gamma(r)$ is fixed by a cyclic group of order $\frac{1}{2}(q+1)$ acting on two orbits of size $\frac{1}{2}(q+1)$. Hence, $q \leq 5$.)

The index [K:K'] = 1 or 2. If it is 2, then K contains PSL(2,p) as a normal subgroup. However, the full normalizer of PSL(2,p) in the symmetric group S_{p+1} is PGL(2,p). We cannot have K=PGL(2,p), since we have already determined all biplanes with Property A. Thus, we conclude that the full block stabilizer must be PSL(2,p).

Finally, if the full automorphism group is larger than the full block stabilizer of B (which is transitive on blocks distinct from B) then it must be transitive on blocks. However a proof may be extracted from [1] that a transitive biplane with full block stabilizer PSL(2,p) has at most 5 points on a block.

Thus, we have shown that if any biplanes exist with Property B with $q \equiv 1 \pmod 4$ and $q>5$, we must have q a prime and PSL(2,q) acting as the full automorphism group. Theorems 3 and 4 are now proven.

6. A Closing Remark. It would be nice to exclude any further examples in this last case. With some patience and sophistication, it might be possible to do this by using some algebraic geometry, in the spirit of Section 2. In this case the appropriate subgroup H stabilizing a graph acts as a dihedral group of order (q+1) on two $\frac{1}{2}(q+1)$-gons. Edges in the first $\frac{1}{2}(q+1)$-gon are of the form $\{x,\sigma(x)\}$ for a fixed element σ; those in the second are of the form $\{x,\tau(x)\}$ for a possibly different element τ. (Both σ and τ are powers of a generator of the cyclic subgroup of index 2 in H.) It should be possible to phrase the intersection properties as a statement about solutions to a system of equations in several variables and to exploit Weil's conjecture, and its generalizations, about the number of points on an algebraic variety in spaces of dimension greater than 2.

BIBLIOGRAPHY

1. P.J.Cameron, "Biplanes", Math. Z. 131(1973) 85-101.

2. P.J.Cameron, "Characterisations of Some Steiner Systems, Parallelisms and Biplanes", Math. Z. 136(1974) 31-39.

3. P.J.Cameron and J.H.Van Lint, Graph Theory, Coding Theory and Block Designs, London Math. Soc. Lecture Notes Series No. 19, London: Cambridge Univ. Press, 1975.

4. L.E.Dickson, Linear Groups, with an exposition of the Galois Field Theory, New York: Dover, 1959 (reprint).

5. M.Hall,Jr., "Symmetric block designs with $\lambda=2$", in Combinatorial Mathematics and its Applications, (proc. Conf. Univ. North Carolina, 1967), Univ. North Carolina Press, 1969, 175-186.

6. J. Hirschfeld, Projective Geometries over Finite Fields, Oxford: Clarendon Press, 1979.

7. W.M.Kantor, "2-transitive symmetric designs", Trans. AMS 146(1969) 1-28.

8. G.S.Kelly, "On Automorphisms of Symmetric 2-Designs", Ph.D.

thesis, Westfield College, University of London, 1979.

9. E.S.Lander, "On Self-Dual Codes and Symmetric Designs", J. London Math. Soc. (to appear).

10. H.Lesbesgue, "Sur l'impossibilite en nombres entiers de l'equation $x^m=y^2+1$", Nouv. Ann. 9(1850) 178-181.

11. H.Lüneburg, "Charakterisierung der endlichen desarguesschen projektiven ebenen", Math. Z. 85(1964) 419-450.

12. P.M.Neumann. "Transitive permutation groups of prime degree, IV: A problem of Mathieu and a theorem of Ito", Proc. London Math. Soc., (3)32(1976) 52-62.

13. B. Puttaswamiah and J.D.Dixon, Modular Representations of Finite Groups, New York:Academic Press, 1977.

14. H.Ryser, Combinatorial Mathematics, Carus Mathematical Monographs 14, New York: Wiley, 1963.

15. J.C.D.S.Yaqub, "On two theorems of Lüneburg", Arch. Math. 17(1966) 485-488.

Ein einfacher Beweis für den Satz von Zsigmondy über primitive Primteiler von $a^N - 1$

Heinz Lüneburg

Der in der Oberschrift angesprochene Satz von Zsigmondy besagt, daß es zu zwei von 1 verschiedenen natürlichen Zahlen a und n stets eine Primzahl p gibt, so daß die Ordnung $\text{ord}_p(a)$ von a modulo p gleich n ist, es sei denn, es ist $n = 2$ und $a + 1$ ist eine Potenz von 2 oder aber es ist $n = 6$ und $a = 2$. Dieser Satz spielt in der Theorie der endlichen, linearen Gruppen eine prominente Rolle. Ferner kann man ihn dazu benutzen zu zeigen, daß es zu jeder natürlichen Zahl n unendlich viele Primzahlen $\equiv 1 \bmod n$ gibt, wie dies schon Zsigmondy tat. Wedderburn benutzte ihn, um seinen berühmten Satz über die Kommutativität endlicher Schiefkörper zu beweisen, der wiederum zur Folge hat, daß jeder endliche projektive Raum und jede endliche desarguessche projektive Ebene pappossch ist. Man kann ihn ebenfalls mit Erfolg dazu benutzen, irreduzible Polynome vom Grad n über GF(a) zu berechnen. In diesem Falle muß a natürlich Potenz einer Primzahl sein.

Für diesen Satz von Zsigmondy gibt es eine Reihe von Beweisen in der Literatur. Die mir bekannten finden sich in [1, 2, 3, 4, 7]. Das Studium der Heringschen Arbeit [4] führte mich zu dem hier wiedergegebenen Beweis, der, wie ich dann feststellte, dem Beweis von Dickson, wie auch dem Beweis von Birkhoff & Vandiver verwandt ist. Da er mir jedoch noch einfacher erscheint als jene, scheint mir seine Publikation gerechtfertigt.

Der hier wiederzugebende Beweis macht wie alle mir bekannten Beweise Gebrauch von den Kreisteilungspolynomen Φ_d, die sich hier in natürlicher Weise anbieten, da die fraglichen Primteiler unter den Primteilern von $\Phi_n(a)$ zu suchen sind. Aus $a^n - 1 = \prod_{d|n} \Phi_d(a)$, wobei das Produkt über alle Teiler d von n zu erstrecken ist, und $a - 1 \geq 1$ für alle von 1 verschiedenen natürlichen Zahlen a folgt $\Phi_n(a) \in \mathbb{N}$ für alle diese a und alle n, da Φ_n ja ein Polynom über $\mathbb{Z}$ ist. Von dieser Bemerkung werden wir gelegentlich Gebrauch machen, ohne sie explizit zu erwähnen.

SATZ 1. *Es seien* a *und* n *von* 1 *verschiedene natürliche Zahlen. Ferner sei* p *eine* $\Phi_n(a)$ *teilende Primzahl. Schließlich sei* $f = \mathrm{ord}_p(a)$. *Es gibt dann eine nicht negative ganze Zahl* i *mit* $n = fp^i$. *Ist* $i > 0$, *so ist* p *die größte* n *teilende Primzahl. Ist* $i > 0$ *und ist* p^2 *ein Teiler von* $\Phi_n(a)$, *so ist* $n = p = 2$.

Beweis. Weil $\Phi_n(a)$ ein Teiler von $a^n - 1$ ist, ist p ein Teiler von $a^n - 1$, so daß f ein Teiler von n ist. (S. etwa [5, Satz I.2.9 a), S. 7].) Es gibt also $i, w \in \mathbb{N}_0$ mit $n = fp^iw$, so daß p kein Teiler von w ist. Setze $r = fp^i$. Dann ist $a^r - 1 \equiv 0 \bmod p$, da f ja ein Teiler von r ist. Ferner folgt

$$(a^n - 1)/(a^r - 1) = (((a^r - 1) + 1)^w - 1)/(a^r - 1) = $$
$$= w + \sum_{i=2}^{w} \binom{w}{i}(a^r - 1)^{i-1}.$$

Dies impliziert $(a^n - 1)/(a^r - 1) \equiv w \not\equiv 0 \bmod p$. Wäre $w > 1$, dh. $r < n$, so wäre $\Phi_n(a)$ ein Teiler von $(a^n - 1)/(a^r - 1)$ und es folgt der Widerspruch $w \equiv 0 \bmod p$. Also ist $w = 1$ und somit $n = fp^i$.

Es sei nun $i > 0$. Weil $f = \mathrm{ord}_p(a)$ nach dem Satz von Euler-Fermat ein Teiler von $p - 1$ ist, folgt, daß p die größte, n teilende Primzahl ist.

Ist weiterhin $i > 0$ und setzt man $s = fp^{i-1}$, so ist p ein Teiler von $a^s - 1$ und es folgt

$$(a^n - 1)/(a^s - 1) = (((a^s - 1) + 1)^p - 1)/(a^s - 1) = $$
$$= p + \tfrac{1}{2}p(p - 1)(a^s - 1) + \sum_{i=3}^{p} \binom{p}{i}(a^s - 1)^{i-1}.$$

Ist $p \geq 3$, so folgt hieraus $(a^n - 1)/(a^s - 1) \equiv p \bmod p^2$. Wegen $(a^n - 1)/(a^s - 1) \equiv 0 \bmod \Phi_n(a)$ ist daher p^2 in diesem Falle kein Teiler von $\Phi_n(a)$. Ist also p^2 ein Teiler von $\Phi_n(a)$, so ist $p = 2$. In diesem Falle ist $f = 1$, dh. $n = 2^i$. Dann ist aber $\Phi_n(a) = a^{2^{i-1}} + 1$. Wäre $i > 1$, so wäre $a^{2^{i-1}} \equiv 1 \bmod 4$ und somit $\Phi_n(a) \equiv 2 \bmod 4$. Also ist $i = 1$, dh. $n = 2$, q. e. d.

Nach diesem Satz kann $\mathrm{ggT}(\Phi_n(a),n)$ weder durch zwei verschiedene Primzahlen noch durch das Quadrat einer Primzahl teilbar sein. Daher gilt das

KOROLLAR. *Sind* a *und* n *von* 1 *verschiedene natürliche Zahlen und ist* $p = \mathrm{ggT}(\Phi_n(a),n) \neq 1$, *so ist* p *die größte* n *teilende Primzahl.*

SATZ 2. *Es seien a und n von 1 verschiedene natürliche Zahlen. Ferner sei* $p = ggT(\Phi_n(a),n)$. *Für* $p = 1$ *sei* $j = 0$ *und für* $p \neq 1$ *sei* p^j *die höchste,* $\Phi_n(a)$ *teilende Potenz von p. Schließlich sei* $\Phi_n^*(a) = p^{-j}\Phi_n(a)$. *Genau dann gilt* $\Phi_n^*(a) = 1$, *wenn entweder* $n = 2$ *und* $a + 1$ *eine Potenz von 2 oder wenn* $n = 6$ *und* $a = 2$ *ist.*

Beweis. Es sei zunächst $n = 2$. Dann ist $\Phi_n(a) = a + 1$. Hieraus folgt, daß genau dann $\Phi_n^*(a) = 1$ ist, wenn $a + 1$ eine Potenz von 2 ist. Es sei also $n \geq 3$. Wegen $\Phi_6 = x^2 - x + 1$ ist $\Phi_6(2) = 3$ und daher $\Phi_6^*(2) = 1$. Es sei schließlich $\Phi_n^*(a) = 1$. Ist ζ eine primitive n-te Einheitswurzel, so ist $|a - \zeta| > a - 1$ und folglich

$$\Phi_n(a) = |\Phi_n(a)| = \prod_\zeta |a - \zeta| > (a - 1)^{\varphi(n)}.$$

Daher ist $\Phi_n(a) > \Phi_n^*(a)$. Also ist $p = ggT(\Phi_n(a),n) \neq 1$. Wegen $n \geq 3$ folgt aus Satz 1, daß $\Phi_n(a) = p\Phi_n^*(a) = p$ ist. Weil p ein Teiler von n ist, ist $p - 1$ ein Teiler von $\varphi(n)$. Somit gilt

$$p = \Phi_n(a) > (a - 1)^{\varphi(n)} \geq (a - 1)^{p-1}.$$

Hieraus folgt $a = 2$. Wäre p^2 ein Teiler von n, so folgte $\Phi_n = \Phi_{n/p}(x^p)$. (S. etwa [6, Satz 14.3, S. 58].) Weiter folgte

$$p = \Phi_n(2) = \Phi_{n/p}(2^p) > (2^p - 1)^{\varphi(n/p)} \geq (2^p - 1)^{p-1}.$$

Dies ist aber unmöglich. Also ist $n = fp$, wobei f ein Teiler von $p - 1$ ist, so daß, wie wir schon feststellten, $f < p$ ist und überdies f und p teilerfremd sind. Nun ist $\Phi_f\Phi_n = \Phi_f\Phi_{fp} = \Phi_f(x^p)$. (S. etwa [6, Satz 14.1o, S. 60].) Daher gilt

$$p(2^p - 1) > p(2^f - 1) \geq \Phi_n(2)\Phi_f(2) = \Phi_f(2^p) > (2^p - 1)^{\varphi(f)}.$$

Also ist $p > (2^p - 1)^{\varphi(f)-1}$ und folglich $\varphi(f) = 1$. Hieraus folgt $f = 1$ oder $f = 2$, so daß $n = p$ oder $n = 2p$ ist. Wäre $n = p$, so folgte $p = \sum_{i=0}^{p-1} 2^i = 2^p - 1$, ein Widerspruch. Also ist $n = 2p$ und daher $p = \sum_{i=0}^{p-1} (-1)^i 2^i = (2^p + 1)/(2 + 1)$, was $p = 3$ zur Folge hat, q. e. d.

Hieraus folgt nun unmittelbar das eingangs zitierte Resultat von Zsigmondy, da es genau dann keinen Primteiler p von $a^n - 1$ mit $ord_p(a) = n$

gibt, wenn $\Phi_n^*(a) = 1$ ist.

Literaturverzeichnis

[1] E. Artin, The orders of the linear groups. Comm. Pure & Appl. Math. 8, 355-365 (1955)

[2] G. D. Birkhoff & H. S. Vandiver, On the integral divisors of $a^n - b^n$. Ann. of Math. (ser. 2) 5, 173-180 (1904)

[3] L. E. Dickson, On the cyclotomic function. Am. Math. Monthly 12, 86-89 (1905)

[4] Ch. Hering, Transitive linear groups and linear groups which contain irreducible subgroups of prime order. Geom. Ded. 2, 425-460 (1974)

[5] B. Huppert, Endliche Gruppen I. Berlin-Heidelberg-New York 1967

[6] H. Lüneburg, Galoisfelder, Kreisteilungskörper und Schieberegisterfolgen. Mannheim-Wien-Zürich 1979

[7] K. Zsigmondy, Zur Theorie der Potenzreste. Monatshefte für Math. & Phys. 3, 265-284 (1892)

Anschrift des Autors

FB Mathematik der Universität
Pfaffenbergstraße 95

D-6750 Kaiserslautern

ON A CLASS OF EDGE-REGULAR GRAPHS

A. Neumaier
Institut f. Angewandte Mathematik
Freiburg, W.-Germany

Summary

A regular graph is 4-regular if any two vertices at distance one resp. two have exactly λ resp. μ common neighbours. Some parameter relations are proved, and certain extremal cases of 4-regular graphs are classified. In particular, we obtain a characterization of double covers of complete graphs related to regular twographs.

All our graphs are finite, connected, undirected, without loops or multiple edges. We call the vertices of a graph points. The neighbourhood $\Gamma(x)$ of a point x of a graph Γ is the graph induced on the points adjacent with x; similarly $\Gamma_i(x)$ denotes the graph induced on the points at distance i from $x \in \Gamma$. A clique is a complete subgraph.

A graph Γ is edge-regular with parameters v, k, λ if there are v points, every point is adjacent with exactly k other points, and every edge is in exactly λ triangles. Γ is called 4-regular if it is edge-regular, and any two points at distance 2 have μ common neighbours (cf. Neumaier [8] for the general concept of t-regularity). The 4-regular graphs of diameter ≤ 2 are just the strongly regular graphs; they are well studied (see e.g. [5], [6], [10]). We call a 4-regular graph proper if its diameter is > 2.

Proposition 1
In a 4-regular graph, the number ℓ of points at distance two from a point x is independent of x and satisfies the relation

$$\ell\mu = k(k-1-\lambda) \tag{1}$$

Proof. Count in two ways the number of paths of length 2. □

Example 1. Regular graphs of girth ≥ 5 ($\lambda=0$, $\mu=1$), and their line graphs ($k=2\lambda+2$, $\mu=1$). Note that for any 4-regular graph $\ell \leq k(k-1)$ with equality iff $\lambda = 0$, $\mu = 1$. More generally, point graphs and line graphs of 1-designs of girth ≥ 5; if there are K points on a line, and R lines through a point then the point graph has $\mu = 1$, $\lambda = K-2$, $k = R(K-1)$, and the line graph has $\mu = 1$, $\lambda = R-2$, $k = K(R-1)$. It is easy to see that every 4-regular graph with $\mu = 1$ arises in this way.

Example 2. The incidence graphs of semisymmetric designs (i.e. square designs where two points are in 0 or Λ blocks, and two blocks have 0 or Λ common points, see Wild [13]); they have $\lambda = 0$, $\mu = \Lambda$. The incidence graphs of symmetric 2-designs, symmetric nets, and partial Λ-geometries arise here. In particular, the symmetric complete bipartite graphs $K_{n,n}$ with a 1-factor deleted appear as the incidence graphs of a degenerate 2-(n,n-1,n-2)-design; here $\lambda = 0$, $\mu = n-2$. (Note that a free construction given by Cameron [3] shows that there are infinitely many infinite examples with $\lambda = 0$ of this type, for each k, μ).

Example 3. 4-regular graphs with $\lambda = 0$, $\mu = 2$ are called rectagraphs by Perkel [9], and are studied in Neumaier [7]. Examples are the n-cube, and the incidence graphs of biplanes. Cameron [4] gave a construction from linear binary codes of minimum weight ≥ 5: Vertices are the cosets, adjacent iff their union contains vectors at Hamming distance one.

Example 4. Every distance-regular graph (Biggs [1], [2]) is 4-regular.

Example 5. A double-cover of a graph Γ^* is a graph Γ such that there is a mapping $*$ of the vertices of Γ to the vertices of Γ^* such that every vertex of Γ^* has exactly two preimages, and $\Gamma(x)^* = \Gamma^*(x^*)$ for all $x \in \Gamma$. It is easy to see that all double covers of complete graphs arise in the following way: Take a graph Γ, a further copy Γ' of Γ, and two further vertices ∞ and ∞'. Add the edges ∞x for $x \in \Gamma$, xy' for $x,y \in \Gamma$, x nonadjacent with y, and $y'\infty'$ for $y \in \Gamma$. This defines a graph Γ^D. If Γ has v vertices then Γ^D is a double cover of K_{v+1}.

Proposition 2
A double cover Γ^D of a complete graph is 4-regular iff Γ is a strongly regular graph with parameters (v,k,λ,μ), where $k = 2\mu$. In this case Γ^D has parameters $k^D = \ell^D = v$, $\lambda^D = k$, $\mu^D = v-1-k$.

Proof. Let Γ^D be 4-regular. Suppose $x \in \Gamma$ has k neighbours in Γ. Then the edge ∞x of Γ^D is in k triangles whence $k = \lambda^D$, and Γ is regular. Let xy be a nonedge of Γ, and suppose that x and y have μ common neighbours in Γ. Then they have $\overline{\lambda} = v-2-2k+\mu$ common nonneighbours in Γ. Hence x and y have $1+\mu+\overline{\lambda} = v-1-2k-2\mu$ common neighbours in Γ^D. Therefore, $v-1-2k+2\mu = \mu^D$. But it is easy to see that $k^D = \ell^D = v$, whence by Prop. 1, $\mu^D = k^D-1-\lambda^D = v-1-k$; Therefore $2\mu = k$. In particular μ is constant.
Finally, let xy be an edge of Γ, and suppose that x and y have λ common neighbours in Γ_o. Then they have $\overline{\mu} = v-2k+\lambda$ common nonneighbours in Γ_o, i.e. xy is in $1+\lambda+\overline{\mu} = v-2k+2\lambda+1$ triangles of Γ^D. Hence $\lambda = \frac{1}{2}(\lambda^D-v+2k-1)$ is constant. So Γ is strongly regular, and satisfies $k = 2\mu$.
Conversely, if Γ is strongly regular and satisfies $k = 2\mu$ then the well-known relation $k(k-1-\lambda) = \mu(v-1-k)$ gives $\lambda = \frac{1}{2}(3k-1-v)$, and a reversal of the above arguments shows that Γ^D is 4-regular with the stated parameters. □

Note that an empty strongly regular graph (with $k=\lambda=\mu=0$) gives K_{vv} with a 1-factor deleted (cf. Example 2). The other 4-regular double covers of K_n are equivalent with regular twographs; see [11], [12]. In [12], Taylor and Levingston state that the double covers of Proposition 2 are in fact distance regular graphs, and he shows that they are characterized by their parameters. We prove the same characterization under the weaker assumption of 4-regularity:

Theorem 1

The parameters of a proper 4-regular graph Γ satisfy $\ell \geq k$. Equality holds iff Γ is a polygon, or a double cover of a complete graph.

Proof. It is clear that the proper 4-regular polygons and double covers of a complete graph satisfy $\ell = k$.

Conversely, let Γ be a proper 4-regular graph with $\ell \leq k$. We shall show that Γ is a polygon, or a double cover of a complete graph.

Step 1. $k = \ell = \lambda+\mu+1$. Moreover, if xy is an edge then every point of $\Gamma_2(x) \cap \Gamma_1(y)$ is adjacent with every point of $\Gamma_3(x) \cap \Gamma_2(y)$.

For by (1), $\ell \leq k$ implies $k \leq \lambda+\mu+1$. To show equality, take $d \in \Gamma_3(x)$, $y \in \Gamma_1(x) \cap \Gamma_2(d)$. There are $k-1-\lambda$ points in $\Gamma_2(x) \cap \Gamma_1(y)$, but this set contains the $\mu \geq k-1-\lambda$ points of $\Gamma_1(y) \cap \Gamma_1(d)$. Hence the two sets coincide, and $\mu = k-1-\lambda$, i.e. by (1), $\ell = k$. If we let d vary over the points of $\Gamma_3(x) \cap \Gamma_2(y)$ we see that the second assertion is true.

Step 2. If $\mu = 1$ then Γ is a polygon.

If $\mu = 1$ then $\lambda = k-2$ (Step 1!). Pick an edge xy, and denote by z the point adjacent to x but not to y (unique since $\lambda = k-2$). Every neighbour of $x \neq y,z$ is adjacent to both y and z, but y and z have only one common neighbour ($\mu=1$), which must be x. Hence y and z are the only neighbours of x. Therefore, $k=2$, $\lambda=0$, and Γ is a polygon.

Now let us fix two points a,d at distance 3. Define

$$B := \Gamma_1(a) \cap \Gamma_2(d), \quad A := \{x \in \Gamma_3(d) \mid \Gamma_1(x) \cap B \neq \emptyset\},$$

$$C := \Gamma_2(a) \cap \Gamma_1(d), \quad D := \{x \in \Gamma_3(a) \mid \Gamma_1(x) \cap C \neq \emptyset\}.$$

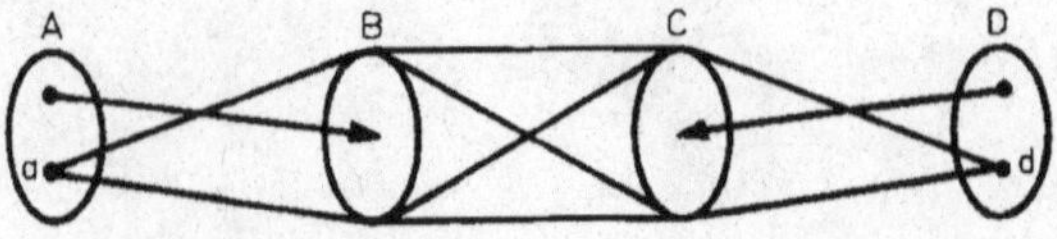

Of course $a \in A$, $d \in D$. The main effort consists in showing that $|A| = 1$, i.e. $A = \{a\}$, and we aim at this now.

Step 3. If $b \in B$ then $\Gamma_1(b) \subseteq A \cup B \cup C$; if $c \in C$ then $\Gamma_1(c) \subseteq B \cup C \cup D$. For if $b \in B$ then $b \in \Gamma_2(d)$; hence $z \in \Gamma_1(b)$ has distance 1,2, or 3 from d. If $z \in \Gamma_1(d)$ then $z \in \Gamma_2(a)$; hence $z \in C$. If $z \in \Gamma_2(d)$ then by Step 1 (for x=a, y=b), z cannot be nonadjcent with a; hence $z \in B$. If $z \in \Gamma_3(d)$ then $z \in A$ since $\Gamma_1(z) \cap B \supseteq \{b\} \neq \emptyset$. So $z \in A \cup B \cup C$. The second statement is symmetric to the first.

Step 4. $A - \{a\} \subseteq \Gamma_1(a)$. –
For if $a' \in A - \{a\}$ but $a' \notin \Gamma_1(a)$ then Step 1 (for x=a, $y \in \Gamma_1(a') \cap B$) gives a contradiction.

Step 5. Each point of $E := A \cup B - \{a\}$ has at most λ neighbours in E; an edge from $A - \{a\}$ to B is in $\lambda-1$ triangles of E. –
By Step 4, $E \subseteq \Gamma_1(a)$, and $\Gamma_1(a)$ has valency λ. This proves the first statement. If $a' \in A - \{a\}$ then $\Gamma_1(a') \cap C = \emptyset$, and if $b \in B \cap \Gamma_1(a')$ then $\Gamma_1(b) \subseteq A \cup B \cup C$. Hence the λ-set $\Gamma_1(a') \cap \Gamma_1(b)$ is in $A \cup B$; so $A \cup B - \{a\}$ contains $\lambda-1$ common neighbours of a' and b.

Step 6. If $a' \in A - \{a\}$ then $\Gamma_1(a) \cap \Gamma_1(a')$ is a λ-subclique of E. –
There is $b \in B \cap \Gamma_1(a')$. Define $A_o = (A-\{a\}) \cap (\Gamma_1(a') \cup \{a'\})$, $B_o = B \cap (\Gamma_1(b) \cup \{b\})$. They are in E. By Step 5, each neighbour of a' or b in E must be a neighbour of both. Hence $|A_o \cup B_o| = \lambda+1$, and a', b are adjacent with all other points of $A_o \cup B_o$. If we replace a' by $a'' \in A_o$ we find similarly that a" is adjacent with all points of $A_o \cup B_o$, and if we replace b by $b' \in B_o$ we find that b' is adjacent with all points of $A_o \cup B_o$. Hence $A_o \cup B_o$ is a clique. But $\Gamma_1(a) \cap \Gamma_1(a') \supseteq A_o \cup B_o \setminus \{a'\}$, and since both sets have size λ, they are the same.

Step 7. If $|A| > 1$ then $\mu \leq \lambda$. –
Take $a' \in A-\{a\}$, $b \in \Gamma_1(a') \cap B$, $c \in \Gamma_1(b) \cap C$. Then a' is adjacent with a and has distance 2 from c. Neighbours of c are in $B \cup C \cup D$, whence $\Gamma_1(a') \cap \Gamma_1(c) \subseteq B \subseteq \Gamma_1(a)$. Hence $\Gamma_1(a') \cap \Gamma_1(c) \subseteq \Gamma_1(a') \cap \Gamma_1(a)$, which implies $\mu \leq \lambda$.

Step 8. If $|A| > 1$ then $\Gamma_1(a)$ is a $(\lambda+1)$-clique. –
$\Gamma_1(a)$ contains the $(\lambda+1)$-clique $(\Gamma_1(a) \cap \Gamma_1(a')) \cup \{a'\}$ constructed in Step 6. Since the valency of $\Gamma_1(a)$ is λ, the $(\lambda+1)$-clique is a component.

If $\Gamma_1(a)$ were not a $(\lambda+1)$-clique, there were another component of valency λ, hence size $\geq \lambda+1 > \mu$ (by Step 7). The size of $\Gamma_1(a)$ would be bigger then $(\lambda+1)+\mu = k$, contradiction.

Step 9. $|A| = 1$. -
Assume that $|A| > 1$. Since a polygon has $|A| = 1$, Γ is not a polygon, hence $\mu > 1$. Choose $c \in C$, and let b, b' be two of the μ points adjacent with a and c. Then $b, b' \in \Gamma_1(a)$, hence by Step 8, they are adjacent and have $\lambda-1$ common neighbours in $\Gamma_1(a)$. But they also have the neighbours $a, c \notin \Gamma_1(a)$, contradiction.

Step 10. If $B \neq \Gamma_1(a)$ then Γ is a polygon. -
Choose $e \in \Gamma_1(a) \setminus B$. Then $\Gamma_1(e) \cap B = \emptyset$ (Step 3, first part) and $\Gamma_1(e) \cap C = \emptyset$ (Step 3, second part). Moreover, for any $b \in B$, $\Gamma_1(b) \subseteq A \cup B \cup C$, whence $\Gamma_1(b) \cap \Gamma_1(e) \subseteq A$. But since $\Gamma_1(e) \cap B = \emptyset$, the left hand side has size $\mu > 0$. Hence by Step 9, $\mu = 1$, and by Step 2, Γ is a polygon.

Step 11. If $B = \Gamma_1(a)$ then Γ is the double cover of a complete graph. -
In this case, by definition of B, $\Gamma_1(a) \subseteq \Gamma_2(d)$. Hence by Step 3, $\Gamma_2(a) \subseteq C \subseteq \Gamma_1(d)$. Since $|\Gamma_2(a)| = \ell = k = |\Gamma_1(d)|$ we have $\Gamma_2(a) = \Gamma_1(d)$. Now a point of $C = \Gamma_2(a) = \Gamma_1(d)$ has μ neighbours in B, λ neighbours in C, and the neighbour d; since $k = \mu+\lambda+1$, there are no further neighbours. Also d has all its neighbours in $\Gamma_1(d) = C$. Hence $\{a\} \cup \Gamma_1(a) \cup \Gamma_2(a) \cup \{d\}$ is a component of Γ, and since Γ is connected, this is Γ. Therefore, Γ contains exactly $1+k+k+1$ points. Hence each point x has a unique point $\bar{x}$ at distance 3 with x. Hence Γ is a double cover of a complete graph K_{k+1} (the mapping $*$ identifies each x with $\bar{x}$). □

A second inequality of Taylor and Levingston [12] for distance regular graphs also generalizes to 4-regular graphs. Again the case of equality can be characterized.

Theorem 2
A proper 4-regular graph Γ satisfies $k \geq 2\lambda+3-\mu$, with equality iff Γ is either the icosahedron ($k=5$, $\lambda=\mu=2$), or the line graph of a regular graph of girth ≥ 5 ($k=2\lambda+2$, $\mu=1$).

Proof. Let xty be an (induced) path of length 2. Denote by q the number of induced quadrangles containing xty, and by c the number of 3-claws

 containing xty. The k-2 vertices $z \neq x,y$ adjacent with t fall into four classes: There are $\mu-1-q$ such vertices adjacent with x and y, $k-2-\lambda-c$ such vertices adjacent with x but not with y, $k-2-\lambda-c$ such vertices z adjacent with y but not with x, and c vertices z adjacent with neither of x,y. Hence $k-2 = (\mu-1-q) + 2(k-2-\lambda-c) + c$, or $k = 2\lambda+3-\mu+q+c$. Hence $k \geq 2\lambda+3-\mu$, and equality implies that Γ contains neither 3-claws nor induced quadrangles.

Now assume that $k = 2\lambda+3-\mu$. If $\mu=1$ then by the remark of Example 1, Γ is the line graph of a regular graph of girth ≥ 5. If $\mu > 1$ then $k \leq 2\lambda+1$. Choose $a \in \Gamma$, and consider $\Gamma(a)$. $\Gamma(a)$ is regular of valency λ, not a clique, and has $k \leq 2\lambda+1$ points. Also, $\Gamma(a)$ contains neither 3-cocliques nor induced quadrangles. Hence the complement of $\Gamma(a)$ is triangle-free, and has no induced subgraph of the shape  ; in particular it contains no induced n-gon with $n \geq 6$. If it contains a pentagon then it is a pentagon (any extra vertex produces a ); otherwise it is easy to see that it is a complete bipartite graph. Hence $\Gamma(a)$ itself is either a pentagon or a union of two cliques. But the latter is impossible since $k \leq 2\lambda+1$. Hence each $\Gamma(a)$ is a pentagon, $k=5$, $\lambda=\mu=2$, and it is a well-known exercise that then Γ is the icosahedron. □

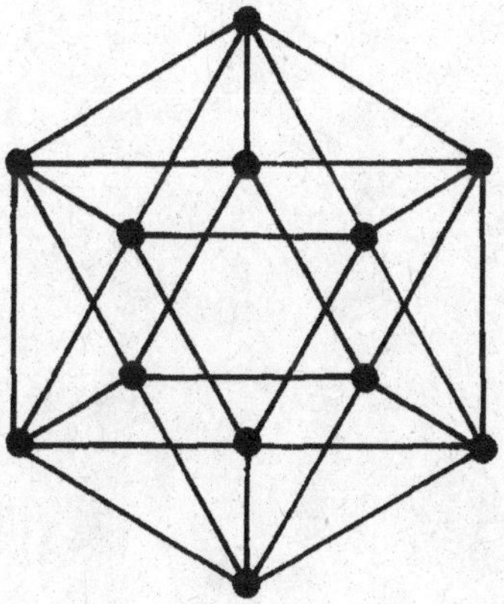

References

1. N.L. Biggs, Algebraic graph theory. Cambridge Univ. Press, 1974.

2. N.L. Biggs, Automorphic graphs and the Krein condition, Geometriae Dedicata 5 (1976), 117-127.

3. P.J. Cameron, Biplanes, Math. Z. 131 (1973), 85-101.

4. P.J. Cameron, personal communication.

5. X. Hubaut, Strongly regular graphs, Discrete Math. 13 (1975), 357-381.

6. A. Neumaier, Strongly regular graphs with smallest eigenvalue -m, Arch. Math. 33 (1979), 392-400.

7. A. Neumaier, Rectagraphs, diagrams, and Suzuki's sporadic simple group, Ann. Discrete Math., to appear.

8. A. Neumaier, Classification of graphs by regularity, J. Combin. Theory B, to appear.

9. M. Perkel, Bounding the valency of polygonal graphs with odd girth, Can. J. Math. 31 (1979), 1307-1321.

10. J.J. Seidel, Strongly regular graphs, an introduction. In: Surveys in Combinatorics, Cambridge Univ. Press 1979, pp. 157-180.

11. D.E. Taylor, Regular twographs, Proc. London Math. Soc. (3) 35 (1977), 257-274.

12. D.E. Taylor and R. Levingston, Distance-regular graphs, Lecture Notes in Mathematics 686, pp. 313-323.

13. P. Wild, Semibiplanes, to appear.

GEOMETRIES UNIQUELY EMBEDDABLE IN PROJECTIVE SPACES

Nicolas PERCSY

Abstract. The search for geometric lattices, which admit an embedding in a projective space satisfying universal or uniqueness properties, has been initiated by W.M. Kantor [2,3] (see also the survey [4]). A recent result of the author [8] leads to a unification and generalization of Kantor's theorems (see Theorem 1). A typical application of this generalization is the following (see Section III.D).

Let G be a geometric lattice which is not the union of two hyperplanes. Assume that each interval [p,1] of G, where p is a point, is isometrically embeddable in PG(m-1,K) for some field K, and that all embeddings of [L,1] in PG(m-2,K), where L is a line, are projectively equivalent, then G is isometrically embeddable in PG(m,K) (and its embeddings satisfy some uniqueness properties).

I. INTRODUCTION

This paper presents various sufficient conditions under which a geometry is embeddable in some projective space.

The geometries considered here are essentially matroids or geometric lattices (actually, they need not be atomic and their dimension may be infinite - see Section II.A). The embeddings are isometric in W.M. Kantor's sense [3] : they are order-monomorphisms of lattices preserving dimension of non-maximal elements ; in the language of matroids, they are restrictions of truncations - see Section II.C. The projective spaces in which we embed may be generalized projective spaces of possibly infinite dimension (see II.B).

The embeddability conditions involved in our results are mainly inspired by Kantor's hypotheses [2,3] : if certain "small" top intervals of a geometry are embeddable, and if their embeddings satisfy some additional properties (mentioned below), then the whole geometry is embeddable.

The emphasis on top intervals can be motivated by the so-called scum theorem (see for instance Welsh [10,p. 324]): this states roughly that any subgeometry of an interval of a geometry G also appears as a subgeometry of some top interval. Hence, by checking top intervals, we are sure that G contains no "forbidden" subgeometry that would make it non embeddable (as, for instance, a non-Desarguesian configuration).

Kantor's additional properties for the embeddings of small intervals are of two kinds. [2] deals with strong embeddings, i.e.

embeddings $i:\underline{G}\to\underline{P}$ of a geometry $\underline{G}$ into a projective space $\underline{P}$ such that $i(\underline{G})$ is a "large" subgeometry of $\underline{P}$. In [3], embeddings are required to satisfy a universal property : an embedding $i:\underline{G}\to\underline{P}$ is "universal" (is a so-called "envelope") if for any other embedding $i':\underline{G}\to\underline{P}'$ of $\underline{G}$, there is an embedding $j:\underline{P}\to\underline{P}'$ such that $ij = i'$.

The purpose of this paper is to continue Kantor's work in several directions. Indeed, some restrictions appear in his deep results : the strong embedding theorem [2] requires that all top rank 3 intervals of the geometries are projective or affine planes ; the universal embedding theorem [3] holds only for embeddings in the class of all projective spaces over a given field ; this field may have no proper subfield isomorphic to itself and the geometries considered are finite-dimensional geometric lattices which are "thick" enough (in the sense of K-rigidity in [3]). In order to get rid of these restrictions, it was necessary to find a new proof of [2,3]. This was made possible, on the one hand, by using a necessary and sufficient embeddability condition obtained by the author [8] and, on the other hand, by restating Kantor's theorems in a form which is more suitable for generalization.

This new statement is based on the following remark : both strong and universal embeddings are unique up to isomorphism in the class of all strong embeddings and the class of embeddings in projective spaces over a given field respectively. This leads naturally to our main "abstract" theorem (Section III, Theorem 1), asserting the embeddability of geometries in which certain top intervals are uniquely embeddable in some class (indeed a category) of embeddings. Actually the class of embeddings is arbitrary : any particular choice of it provides a particular "concrete" embedding theorem. For instance, Kantor's results can be obtained by considering the above-mentioned classes (see IV.A,B). Other applications, based on other choices, are given in IV.C,D ; they show how the restricting hypotheses in [2,3] can be dropped. While the proof of Theorem 1 is given in Section III, the proof of the applications could only by sketched in view of the length allowed for this paper (for more details and new applications, see [9]).

As a conclusion, Theorem 1 provides a unique proof of both main embedding theorems in [2] and [3], allows to weaken their statements (see IV.A,B), holds for a larger class of geometries (not necessarily atomic, nor finite-dimensional) and opens the way to new applications. Let us note that the crucial result in the proof of Theorem 1 is [8], and that the (elementary) terminology of category theory used in the statement of Theorem 1 is not just a way of formalization. Categories are needed for obtaining the various applications, especially in IV.D as explained there.

II. TERMINOLOGY AND NOTATIONS

A. GEOMETRIES. A lattice $\underline{G}$ with least element 0 and greatest element 1 will be called a *geometry* if it satisfies the following axioms ([1]) :
(F) *for any* $X \in \underline{G} \setminus \{1\}$, *all chains from* 0 *to* X *are finite* ;
(G) $\underline{G}$ *is semimodular* (*i.e. if* X *covers* $X \wedge Y$ *and* $X \ngeq Y$, *then* $X \vee Y$ *covers* Y) ;
(H) *the join of all atoms is* 1.

By (G), for any $X \in \underline{G}$, either all maximal chains from 0 to X have the same number m of elements, or they are all infinite. The *rank of* X is defined to be m-1 in the former case and infinite in the latter ; the *rank of* $\underline{G}$ is the rank of 1. By (F), any $X \in \underline{G} \setminus \{1\}$ has finite rank and $\underline{G}$ has finite or countable infinite rank.

The elements of a geometry are often called *varieties* ; two varieties V,W are *incident* if one of the possibilities $V \leq W$ and $W \leq V$ occurs. Varieties of rank 1,2 and 3 are also called *points*, *lines* and *planes* respectively ; in a geometry of finite rank n, the varieties of rank n-1 and n-2 are *hyperplanes* and *colines*.

Let us note that (H) does not mean that any geometry is an atomic lattice (point lattice) : the class of *atomic geometries* of finite rank coincides with the class of geometric lattices (or lattices of flats of a matroid, see for instance Aigner [1]).

Let $\underline{G}$ be a geometry of (possibly infinite) rank n and $V \in \underline{G} \setminus \{1\}$ a rank m variety. The interval $\{X \in \underline{G} : X \geq V\}$ is obviously a geometry, sometimes called *residual* or *local geometry at* V ; it will be denoted by $\underline{G}_V$. If n is finite, $\underline{G}_V$ has rank n-m. The *rank r truncation of* $\underline{G}$, where $1 \leq r \leq n$, is the geometry of all varieties of $\underline{G}$ whose rank is less than r, together with 1.

B. PROJECTIVE GEOMETRIES. The set of all finite rank elements of an atomic modular lattice $\underline{M}$, together with 1, constitutes a geometry. This geometry, and all its truncations, are called *projective geome-*

([1]) We find it clearer to add the axiom (H) (not used in the definition of a geometry in Percsy [8]) in order to lighten the statements of the results below.

tries (²). If $\underline{M}$ is the subspace lattice of an arbitrary dimensional vector space over a skewfield K, then any projective geometry obtained from $\underline{M}$ is called a *projective geometry over* K. In particular, the *notation* PG(K^n) will be used to denote the rank n projective geometry of all subspaces of the standard n-dimensional vector space over K.

C. ISOMETRIES. An *isometry* (³) (Kantor [3]) from a geometry $\underline{G}$ into a geometry $\underline{H}$ is a mapping $i:\underline{G}\to\underline{H}$ satisfying the following conditions :

(I1) *for any* $V,W\in\underline{G}$: $V\leq W$ *in* $\underline{G}$ *if and only if* $V^i\leq W^i$ *in* $\underline{H}$;

(I2) *for any* $X\in\underline{G}\setminus\{1\}$, *the rank of* X *in* $\underline{G}$ *and the rank of* X^i *in* $\underline{H}$ *are equal* ;

(I3) $1^i = 1$.

Isometries are thus order-monomorphisms ; they are also join-preserving (Percsy [5]) :

(I4) *for any family* $\{X_k : k\in I\}$ *of elements of* $\underline{G}$, *if* $\bigvee_{k\in I} X_k \neq 1$, *then* $(\bigvee_{k\in I} X_k)^i = \bigvee_{k\in I} X_k^i$.

The following obvious properties will be used implicitely.

(i) The composite of isometries is an isometry.

(ii) An isometry is surjective if and only if it is an isomorphism of geometries.

(iii) If $i:\underline{G}\to\underline{H}$ is an isometry, then for every $V\in\underline{G}\setminus\{1\}$, the restriction of i to $\underline{G}_V$ is an isometry from $\underline{G}_V$ into $\underline{H}_{i(V)}$.

All restrictions of an isometry will be denoted by the same letter as that isometry.

D. EMBEDDINGS. An *embedding* is a triple $\varepsilon = (\underline{G},\underline{P},i)$, where $\underline{G}$ is an arbitrary geometry, $\underline{P}$ is a projective geometry and i is an isometry from $\underline{G}$ into $\underline{P}$. We say that ε is an *embedding of* $\underline{G}$ and the *rank of* ε is the rank of G. Two embeddings $(\underline{G},\underline{P},i)$ and $(\underline{G},\underline{P}',i')$ of a given geometry G are *equivalent* or *isomorphic* if there is an *equivalence* or *isomorphism* of embeddings between them, i.e. an isomorphism $\alpha:\underline{P}\to\underline{P}'$ of projective geometries such that $i\alpha = i'$.

In view of the unicity properties introduced below, it is useful to think of the class of all embeddings as a *category* E, whose morphisms (actually isomorphisms) are the equivalences defined above.

(²) It follows from wellknown properties (recalled in [2], [6,8]) that a projective geometry is a truncation of the subspace lattice of a generalized projective space.

(³) Contrarily to [6,7,8], the words "embedding" and "isometry" will be used here in different technical meanings, although the underlying notions are the same.

III. MAIN EMBEDDING THEOREM

A. UNIQUE AND RIGID EMBEDDINGS. Let C be a category of embeddings which is not necessarily a full subcategory of E (i.e. all possible equivalences between any two elements of C are not necessarily morphisms of C). A geometry $\underline{G}$ is *C-embeddable* if there is an embedding of $\underline{G}$ that belongs to C. An embedding $\varepsilon = (\underline{G},\underline{P},i)\in C$ is *rigid in* C if the only isomorphism from ε onto ε is the identity (i.e. the only automorphism of $\underline{P}$ that is a morphism of C fixing all elements of $\underline{G}^i$ is the identity). A C-embeddable geometry $\underline{G}$ is called *uniquely C-embeddable* if given any two embeddings $\varepsilon,\varepsilon'\in C$ of $\underline{G}$, there is at least one isomorphism from ε onto ε' belonging to C. Finally, $\underline{G}$ is called *rigidly C-embeddable* if all C-embeddings of $\underline{G}$ are rigid - or equivalently, if for any two C-embeddings ε,ε' of $\underline{G}$, there is at most one C-isomorphism between them.

Clearly, a geometry $\underline{G}$ is *uniquely and rigidly C-embeddable* if it has a rigid C-embedding that is equivalent, in C, to all its other C-embeddings ; in other words, $\underline{G}$ is C-embeddable and there is a unique C-isomorphism between any two C-embeddings ε,ε' of $\underline{G}$.

B. LOCAL CATEGORY. Let C be a category of embeddings of (possibly infinite) rank $n\geq 3$ (not necessarily a full subcategory of E). We define the *residual* or *local category* C' as follows :

(i) the objects of C' are all restrictions $\varepsilon_X = (\underline{G}_X,\underline{P}_{i(X)},i)$ of the embeddings $\varepsilon = (\underline{G},\underline{P},i)\in C$, where X is any rank one variety of $\underline{G}$ (and $\underline{P}_{i(X)}$ denotes the local projective geometry of $\underline{P}$ at $i(X)$, see II.A) ;

(ii) the restriction of any morphism of C from $\varepsilon = (\underline{G},\underline{P},i)\in C$ onto $\eta = (\underline{G},\underline{Q},j)\in C$ is a morphism of C' from ε_X onto η_X ; also all composites of such restrictions will be morphisms of C'.

It is worth mentioning that, in general, the composite of restrictions considered in (ii) is not such a restriction : the following counter-example actually occurs in the proof of Theorem 1. Let $\varepsilon = (\underline{G},\underline{P},i)$, $\varepsilon' = (\underline{G}',\underline{P},i')$, $\eta = (\underline{G},\underline{Q},j)$, $\eta' = (\underline{G}',\underline{Q}',j')$ be embeddings of a category C ; let α, $\alpha'\in C$ be equivalences from ε onto η, and from η' onto ε' respectively. Assume that, for some rank one varieties $X\in\underline{G}$, $X'\in\underline{G}'$, we have $\underline{G}_X = \underline{G}'_{X'}$ and $\varepsilon_X = \varepsilon'_{X'}$. Then α,α' are morphisms of C' ; their composite is an equivalence from $\eta'_{X'}$ onto η_X, which is not necessarily the restriction of an equivalence of C.

C. THEOREM 1. *Let $\underline{G}$ be a rank n geometry, where $n\geq 5$ may be infinite. Let C_1,C_2,C_3 be arbitrary categories of embeddings of rank n-1,n-2,n-3 respectively, such that $C_2\supset C_1'$ and $C_3\supset C_2'$. Assume that, for any point,* *line or plane* $X \in \underline{G}$, $\underline{G}_X$ *is* C_1*-embeddable, uniquely* C_2*-embeddable, rigidly* C_3*-embeddable respectively.*

Then there is an embedding $(\underline{G}, \underline{P}, i)$ *of* $\underline{G}$ *such that, if* $X \in \underline{G}$ *has rank r, for r=1,2,3, then* $(\underline{G}_X, \underline{P}_{i(X)}, i)$ *is equivalent to some embedding of* $\underline{G}_X$ *belonging to* C_r.

Proof. We shall apply the necessary and sufficient embeddability condition of [8, théorème 3.1]. For every point $p \in \underline{G}$, let us choose an embedding $(\underline{G}_p, \underline{P}(p), \hat{p}) \in C_1$; thus axiom (L) of [8] is satisfied. If p,q are distinct points of G, and if L denotes the line $p \vee q$, then G_L is uniquely C_2-embeddable ; as a consequence, since $C_2 \supset C_1'$, the embeddings $(\underline{G}_L, \underline{P}(p)_{\hat{p}(L)}, \hat{p})$ and $(\underline{G}_L, \underline{P}(q)_{\hat{q}(L)}, \hat{q})$ are equivalent in C_2. Hence, by Lemma 4.4 of [7], there is a family $\{\alpha_{pq} : p,q$ are distinct points of $\underline{G}\}$ of isomorphisms of projective geometries such that, for any three distinct points $p,q,r \in \underline{G}$:

(i) α_{pq} is an equivalence (belonging to C_2) from $(\underline{G}_L, \underline{P}(p)_{\hat{p}(L)}, \hat{p})$ onto $(\underline{G}_L, \underline{P}(q)_{\hat{q}(L)}, \hat{q})$ (where $L = p \vee q$) ;

(ii) $\alpha_{pq} = \alpha_{qp}^{-1}$;

(iii) if p,q,r are on a common line L, then $\alpha_{pq}\alpha_{qr}$ and α_{pr} coincide on $\underline{P}(p)_{\hat{p}(L)}$.

Now [8] applies if we can prove that (iii) holds when $p \vee q \vee r$ is a plane N. Clearly, $\alpha_{pq}\alpha_{qr}$ and α_{pr} are both equivalences of the embeddings $(\underline{G}_N, \underline{P}(p)_{\hat{p}(N)}, \hat{p}) \in C_3$ and $(\underline{G}_N, \underline{P}(r)_{\hat{r}(N)}, \hat{r}) \in C_3$. Since $\underline{G}_N$ is rigidly C_3-embeddable, these equivalences must be equal on $\underline{P}(p)_{\hat{p}(N)}$. The existence of an embedding $(\underline{G}, \underline{P}, i)$ is thus proved. The final statement of the theorem follows immediately from [8, 3.1, (T)].

IV. APPLICATIONS

This section is devoted to some embedding theorems, providing sufficient embeddability conditions, which can be deduced from Theorem 1. We first show (Sections A and B) that Theorem 1 provides a unified proof of W.M. Kantor's universal embedding and strong embedding theorems [2,3]. Therefore, we state the main result of [3] in an equivalent form (involving uniqueness and rigidity as defined in III.A) ; this new version may be simpler to use sometimes and enables us to get generalizations (Section III.C and [9]). Finally, Section III.D contains a new kind of application of Theorem 1.

A. KANTOR'S UNIVERSAL EMBEDDING THEOREM. Kantor [3] deals with embeddings in projective spaces over a given field. Let K be a (not necessarily commutative) field and let us denote by C(K) the full subcategory of E of all embeddings of any finite rank geometry in a finite-dimensional projective space over K. A *K-envelope* [3] is an embedding $(\underline{G},\underline{P},i)\in C(K)$ satisfying the following universal property : for any embedding $(\underline{G},\underline{Q},j)\in C(K)$, there is a unique isometry $h:\underline{P}\to\underline{Q}$ such that $ih = j$. A geometry $\underline{G}$ is *K-rigid* [3] if every embedding $(\underline{G},\underline{P},i)\in C(K)$ of $\underline{G}$, such that $\bigvee(i(\underline{G})\setminus\{1\}) = \underline{P}$, is rigid in C(K) in the sense of III.A.

THEOREM 2. (Kantor [3]).
*(#) Let K be a field isomorphic to no proper subfield of itself.
An atomic geometry $\underline{G}$ of finite rank $r \geq 5$ has a K-envelope if for some integer d, with $1\leq d\leq r-4$, $\underline{G}_X$ has a K-envelope (resp. is K-rigid) whenever $X\in\underline{G}$ has rank d or d+1 (resp. d+2).*

In order to prove that this result is a corollary of Theorem 1, we shall state it in a different, slightly weaker and more general form. A geometry is said to be *uniquely embeddable* in $PG(K^n)$ (resp. to have *a rigid embedding* in $PG(K^n)$), if it is uniquely embeddable (resp. it has a rigid embedding) in the category of all embeddings in $PG(K^n)$. It follows from [3,Lemma 5 and Corollary] that an atomic geometry of finite rank ≥ 3 has a K-envelope if and only if it is uniquely and rigidly embeddable in some $PG(K^n)$. Hence, Theorem 2 is contained in the following one.

THEOREM 3. *Assume that hypothesis (#) of Theorem 2 holds.
Let $\underline{G}$ be a geometry (not necessarily atomic) of finite rank $r \geq 5$, and let n,d be integers with $r\leq n$ and $1\leq d\leq r-4$; assume that any rank d+i element $X\in\underline{G}$, where i = 0,1 or 2, satisfies condition (i) below :
(0) $\underline{G}_X$ is embeddable in $PG(K^{n-d})$;
(1) $\underline{G}_X$ is uniquely embeddable in $PG(K^{n-d-1})$;
(2) $\underline{G}_X$ has a rigid embedding in $PG(K^{n-d-2})$.
Then $\underline{G}$ is uniquely and rigidly embeddable in $PG(K^n)$.*

Sketch of proof. We use induction on d.
If d = 1, the embeddability of $\underline{G}$ follows from Theorem 1 by considering as C_k the category $C(K^{n-k})$ of all embeddings in projective spaces isomorphic to $PG(K^{n-k})$, with k=1,2,3.
The induction is based on a lemma, similar to [3,Theorem 2], which states that if hypotheses (0),(1),(2) hold for some d>1, they hold also

for d' = d-1. The uniquess and rigidity of the embedding of $\underline{G}$ is proved in the same way.

B. KANTOR'S STRONG EMBEDDING THEOREM. One of the main theorems of [2,Theorem 2] is concerned with strong embeddings : i.e. embeddings $(\underline{G},\underline{P},i)$ such that $i(\underline{G})$ is so "large" in $\underline{P}$ that many varieties of $\underline{P}$, which are not in $i(\underline{G})$, can be obtained as intersections (in $\underline{P}$) of elements of $i(\underline{G})$. Let us show that Kantor's strong embedding is a consequence of Theorem 1.

Let $\underline{G}$ be an atomic geometry of finite rank $r \geq 3$ and $\underline{P}$ a projective geometry which is not a proper truncation of another projective geometry (see II.B). We define a *strong embedding* [2] $(\underline{G},\underline{P},i)$ by induction on the rank r of $\underline{G}$:

(1) When r = 3, $(\underline{G},\underline{P},i)$ is strong if either $i(\underline{G}) = \underline{P}$ or $i(\underline{G})$ is an affine plane obtained from $\underline{P}$ by removing one of its lines.

(2) When r>3, $(\underline{G},\underline{P},i)$ is strong if $(\underline{G}_p,\underline{P}_{i(p)},i)$ is strong for all points $p \in G$ and each point and line $A \in \underline{P} \setminus i(\underline{G})$ is the intersection (in $\underline{P}$) of some family of elements of $i(\underline{G})$.

Let us note that for any strong embedding $(\underline{G},\underline{P},i)$, $\underline{G}$ and $\underline{P}$ have the same rank.

THEOREM 4. (Kantor [2,Theorem 2]) *Let* $\underline{G}$ *be an atomic geometry of finite rank* $r \geq 5$ *and d an integer such that* $1 \leq d \leq r-4$. *Assume that for all rank d elements* $X \in \underline{G}$, *the local geometry* $\underline{G}_X$ *has a strong embedding ; then* $\underline{G}$ *has a strong embedding.*

Proof. We apply Theorem 1 by choosing $C_1 = C_2 = C_3 = CS$ the category of all strong embeddings:each element of CS is uniquely and rigidly embeddable in it as follows from [2,(UL6),(UL4) p. 182].

A notion of strong embedding can be defined for a more general class of geometries, essentially by weakening part (1) of the above definition : hence Theorem 4 can be extended (see [9] for more details).

C. GENERALIZING SECTION A TO ALL FIELDS AND ARBITRARY DIMENSION.

Theorems 2 and 3 of Section A are stated with two restricting hypotheses :

(1) the hypothesis (#) on the field ;

(2) the finiteness of the rank of both the geometry and the projective space.

While Kantor's proof of Theorem 2 cannot be extended to the cases where (1) or (2) does not hold, our proof of Theorem 3 (containing Theorem 2) actually uses neither (1), nor (2). Unfortunately, no geometry satisfies the hypotheses of Theorem 3 when (1) or (2) is false ! For in such a case, a projective space can be embedded properly in itself and hence, *no geometry is uniquely embeddable* in such a space.

Indeed, this provides an idea to generalize Theorem 3 by dropping (1) and (2). Let C'(K) be the category of all embeddings ($\underline{G}$,$\underline{P}$,i) such that $\underline{G}$ is any geometry (not necessarily atomic, nor of finite rank), $\underline{P}$ is any projective geometry over K and *there is no projective geometry* $\underline{P}'$, *isomorphic to* $\underline{P}$, *that is properly (isometrically) embedded in* $\underline{P}$ *and contains* $i(\underline{G})$. If we restrict ourselves to those embeddings that belong to C'(K), there will be geometries uniquely embeddable in C'(K).

THEOREM 5. *Let* $\underline{G}$ *be a (not necessarily atomic) geometry of (possibly infinite) rank* $r \geq 5$ *and d an integer such that* $1 \leq d \leq r-4$ *; assume the following :*
(i) for any rank d element $X \in \underline{G}$, *the geometry* $\underline{G}_X$ *has an embedding* $(\underline{G}_X, \underline{P}, i)$ *in some* $\underline{P} = PG(K^m)$ *(m may be infinite) such that* $(\underline{G}_Y, \underline{P}_{i(Y)}, i)$ *belongs to C'(K) whenever* $Y \in \underline{G}$ *contains X and has rank d+1 or d+2 ;*
(ii) for any rank d+1 (resp. d+2) element $Y \in \underline{G}$, *the geometry* $\underline{G}_Y$ *is uniquely embeddable (resp. has a rigid embedding) in the category of all embeddings in* $PG(K^{m-1})$ *(resp.* $PG(K^{m-2})$*) belonging to C'(K).*
Then $\underline{G}$ *is uniquely and rigidly embeddable in the category of all embeddings in* $PG(K^{m+d})$ *that belong to C'(K).*

Proof. The proof is the same as that of Theorem 3, but instead of the category C_k of *all* embeddings in $PG(K^{n-k})$, we consider the *subcategory* of those embeddings that belong to C'(K).

A similar way to avoid difficulties when (1) or (2) is false is as follows : in the category C(K) of all embeddings in projective geometries over K, consider the (full) subcategory $C_R(K)$ of those embeddings which are rigid. Clearly, $C_R(K)$ is contained in C'(K), and we can state another version of Theorem 5 with the following changes :
- in (i), $(\underline{G}_Y, \underline{P}_{i(Y)}, i)$ is assumed to be rigid (instead of beeing in C'(R)) for rank d+1,d+2 elements Y containing X ;
- (ii) requires only that all rigid embeddings of $\underline{G}_Z$ in $PG(K^{n-1})$, where $Z \in \underline{G}$ has rank d+1, are equivalent.

D. WEAKENING RIGIDITY HYPOTHESIS. So far, all applications of Theorem 1 have been deduced from it by considering certain *full* subcategories of the category E of all embeddings. But we are allowed to consider subcategories in which both the embeddings *and the equivalences* are particular. This leads to a generalization of Theorem 3 and 5 in which the rigidity hypothesis is strongly weakened : instead of a rigidity condition, it is only assumed that the geometry is not the union of two hyperplanes.

We need a definition : a geometry $\underline{G}$ is *uniquely projectively embeddable* in $\underline{P} = PG(K^n)$ if any two embeddings of $\underline{G}$ in $\underline{P}$ are equivalent via an equivalence induced by a member of PGL(n,K) (not an arbitrary collineation of $\underline{P}$ as in Sections A and C).

THEOREM 6. *Let $\underline{G}$ be a (not necessarily atomic) geometry of (possibly infinite) rank $r \geq 5$ and d an integer such that $1 \leq d \leq r-4$. Assume that*
(i) for any rank d element $X \in \underline{G}$, the geometry $\underline{G}_X$ has an embedding $(\underline{G}_X, \underline{P}, i)$ in some $\underline{P} = PG(K^m)$; moreover, when one of the conditions (1) and (2) of Section C is false, the embeddings $(\underline{G}_Y, \underline{P}_{i(Y)}, i)$ are required to be in C'(K) for all rank d+1 or d+2 element Y containing X ;
(ii) for any rank d+1 element $Z \in \underline{G}$, the geometry $\underline{G}_Z$ is uniquely projectively embeddable in $PG(K^{m-1})$.
If, furthermore, $\underline{G}$ is not the union of two hyperplanes, then it is (uniquely projectively) embeddable in $PG(K^{m+d})$.

Sketch of proof. Like in Theorem 3 and 5, we work by induction on d. The main differences with the proofs of the previous results are as follows.
(a) In order to apply Theorem 1, we consider the category $CP(K^{m+d})$, whose elements are the embeddings $(\underline{H}, \underline{P}_A, i)$, where $\underline{P}_A$ is a local geometry at some A of $\underline{P} = PG(K^{m+d})$, and whose isomorphisms are induced by an element of PGL(m+d,K). A geometry is thus uniquely embeddable in that category if and only if it is uniquely projectively embeddable in some $PG(K^n)$.
(b) An embedding $(\underline{H}, \underline{Q}, i) \in CP(K^{m+d})$ is rigid in that category if and only if no nontrivial element of $PGL(\underline{Q})$ fixes all points of $i(\underline{H})$. This is certainly realized when no coline of $\underline{H}$ is contained in (at most) two hyperplanes.

Let us conclude with a remark. While categories appear first here as a simple way of formalizing the definitions of uniqueness and rigidity in Section III, they are actually useful and convenient in the statement and proof of the main result in Section III, as well as the proof of its various applications in Section IV. Indeed, they are needed in Section IV.D (where both the kind of embeddings and of equivalences are restricted) : they avoid a new version of Section III, with new (but quite similar) definitions and main result.

REFERENCES

1. Aigner M., *Combinatorial theory*, Springer-Verlag, New York 1979.
2. Kantor W.M., Dimension and embedding theorems for geometric lattices, *J. Combin. Theory Ser.* A 17 (1974), 173-195.
3. Kantor W.M., Envelopes of geometric lattices, *J. Combin. Theory Ser.* A 18 (1975), 12-26.
4. Kantor W.M., Some highly geometric lattices, *Atti Conv. Lincei* 17 (1976), 183-191.
5. Percsy N., *Plongement de géométries*, Thèse de doctorat, ULB, Bruxelles 1980.
6. Percsy N., Embedding geometric lattices in a projective space, *Proc. 2nd Conf. on Finite Geometries and Designs*, Cambridge U. Press, London 1981.
7. Percsy N., Locally embeddable geometries, to appear in *Arch. Math.* (1981).
8. Percsy N., Une condition nécessaire et suffisante de plongeabilité pour les treillis semi-modulaires, to appear in *European J. Combin.* (1981).
9. Percsy N., Sufficient embeddability conditions for geometric lattices, in preparation.
10. Welsh D.J.A., *Matroid theory*, Academic Press, London 1976.

Nicolas PERCSY
Université de l'Etat à Mons
av. du Champ de Mars, 24
B - 7000 - MONS - BELGIUM

STRONG POINT STABLE DESIGNS

Karl Erich Wolff
Hein-Heckroth-Str. 27
D 6300 Gießen

ABSTRACT

A design (incidence structure) D with incidence matrix A is called *point stable*, if $AA^TJ = \alpha J$ (J the all-one-matrix, $\alpha \in \mathbb{N}$). D is called *strong*, if D is a connected regular point stable design with two or three eigenvalues. The most important strong designs are the 2-design the partial geometric designs, (r,λ)-designs, regular point stable semi partial geometric designs, 2-PBIBD's and strongly regular graphs.
The strong designs are characterized as the regular point stable designs whose multigraphs are linear strongly regular. The eigenvalues of strong designs may be expressed in geometrical terms. For any regular point stable design D we determine the "place" of the multigraph of D in the rank classification scheme. The point graph of a strong design with exactly two connection numbers is strongly regular.

1. INTRODUCTION

The characterization of connected regular graphs (by HOFFMAN [5]) and of partial geometric designs (by BOSE, BRIDGES, SHRIKHANDE [2]) and a classification of 1-designs (by NEUMAIER (unpublished)) led the author to the introduction of point stable designs, semi partial geometric designs [7] and to the rank classification of point stable designs [8,9] BOSE, SHRIKHANDE, SINGHI [3] proved that the multigraph of a partial geometric design is strongly regular. We extend this result to strong designs, using the Hoffman-polynomial and the fact that the adjacency matrix of the multigraph of a regular point stable design is regular. Then we show that the eigenvalues of a strong design may be expressed in geometrical terms. Finally we determine for any regular point stable design the "place" of its multigraph in the rank classification scheme.

2. NOTATIONS AND PRELIMINARIES

A design (or incidence structure) $D = (\mathcal{P}, \mathcal{B}, I)$ with incidence matrix A is called *point stable* (*PSI, point stable incidence structure*), if $NJ = JN$, where $N = AA^T$ is the connection matrix of D and J the all-one-matrix. This is equivalent to $NJ = \alpha J$ for some $\alpha \in \mathbb{N}$. In this case D is called a $PSI(\alpha)$. Throughout let $v = |\mathcal{P}|$, $b = |\mathcal{B}|$.
To exclude degenerate cases we shall always assume

(1) $\alpha < \sigma \quad (\sigma := |I|)$,

which is (for connected point stable designs) equivalent (cf.[7]) to the condition, that the number s of nonzero eigenvalues of D satisfies

(2) $s := |\mathrm{spec}N \setminus \{0\}| \geq 2$.

A design with incidence matrix A is called *r-regular*, if $AJ = rJ$. An r-regular $PSI(\alpha)$ is called a $PSI(\alpha,r)$.
The following results (3,4) (cf. [7]) contain the theorem of HOFFMAN [5] for connected regular graphs.

(3) A design with connection matrix N is connected and point stable iff there is a polynomial $f(X) \in \mathbb{R}[X]$ with $f(N) = J$.

(4) If D is a connected $PSI(\alpha)$, then there is a unique polynomial $h(X) \in \mathbb{R}[X]$ of minimal degree satisfying
$h(N) = J$ (the Hoffman-equation),
namely the Hoffman-polynomial $h(X) = v \prod_{\rho \in \mathrm{spec}N \setminus \{\alpha\}} \frac{X - \rho}{\alpha - \rho}$.

The *point rank Pr(D)* of a design D is defined in [8] as the minimal degree of a monic polynomial $f(X) \in \mathbb{R}[X]$ satisfying $f(N) = tJ$ for some $t \in \mathbb{R}$.
If D is a connected PSI with s nonzero eigenvalues, then

(5) $Pr(D) = |\mathrm{spec}N| - 1$,
$Pr(D) = s - 1 \quad \text{iff} \quad \det N > 0$,
$Pr(D) = s \quad \text{iff} \quad \det N = 0$.

<u>Definition</u>: A design D is called *strong*, if D is a connected $PSI(\alpha,r)$ with $|\mathrm{spec}N| \in \{2,3\}$ (i.e. $Pr(D) \in \{1,2\}$).

Using the rank classification theorem (RCT) (cf.[8,9]) we obtain the following rank classification scheme for connected regular point stable designs with $s \in \{2,3\}$, where the class of strong designs is included in fat lines:

The rank classification scheme:

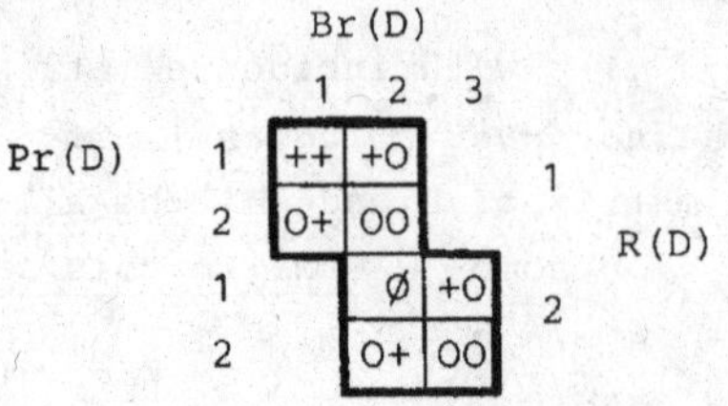

$s = 2$ $(C(D)=1)$

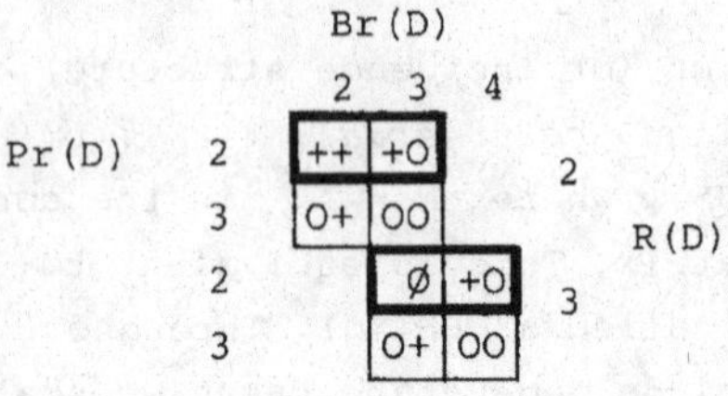

$s = 3$ $(C(D)=2)$

Comments: (i) Example: The sign "+O" means $\det(AA^T)>0$, $\det(A^TA)=0$.
(ii) The *block rank Br(D)* is the point rank of the dual of D.
The *rank R(D)* and the *column rank C(D)* are defined in [8] as the minimal degree of a monic polynomial $f(X) \in \mathbb{R}[X]$ satisfying an equation

(6) $f(N)A = tJ_{v,b}$ for some $t \in \mathbb{R}$

respectively

(7) $f(N)A = J_{v,1}T$ for some $(1\times b)$-matrix T.

(iii) The two empty-set-signs indicate two special cases of the following result (proved in [8]):

(8) If D is a PSI(α,r) with s nonzero eigenvalues, then
$R(D) = s$, $Pr(D) = s-1$ ===> $Br(D) = s+1$.

To describe the most important strong designs we mention the following results for connected point stable designs (cf.[8]):

(9) $R(D) = s-1$ iff D is a 1-design ,
(10) $s = 2$ iff D is semi partial geometric iff $C(D) = 1$,
(11) $R(D) = 1$ iff D is partial geometric ,
(12) $Pr(D) = 1$ iff D is an (r,λ)-design ,
(13) $R(D) = Pr(D) = 1$ iff D is a 2-design ,
(14) $R(D) = Pr(D) = Br(D) = 1$ iff D is a square (v=b) 2-design ,
(15) All 2-PBIBD's with $\det(N)>0$ are strong with $s = 3$ and
$R(D) = Pr(D) = 2$.

3. STRONG DESIGNS AND STRONGLY REGULAR MULTIGRAPHS

In this section we show that the strong designs are just the regular point stable designs whose multigraphs are linear strongly regular.

For any design $D = (P, B, I)$ let $Mg(D)$ denote the multigraph of D, i.e. the multigraph on P, whose connection numbers $[p,q]'$ satisfy

(16) $[p,q]' = [p,q] \qquad (p,q \in P, \ p \neq q)$.

If D is a $PSI(\alpha,r)$ with connection matrix N, then the adjacency matrix H of $Mg(D)$ satisfies

(17) $H = N - rE \qquad (E = \text{identity matrix})$,

(18) $HJ = (\alpha - r)J$,

hence $Mg(D)$ is $(\alpha-r)$-regular.

For any design $D = (P, B, I)$ the number of p-q-paths (p,B,x,C,q) with $x \in P \setminus \{p,q\}$ is

(19) $w(p,q) = \sum_{x \in P \setminus \{p,q\}} [p,x][x,q] = (H^2)_{p,q} \qquad (p,q \in P)$.

BOSE, SHRIKHANDE, SINGHI [3] defined a *strongly regular multigraph* as a regular multigraph with (20,21):

(20) constant loop degree d (for some $d \in \mathbb{N}$), i.e.

$$l(p) := \frac{1}{2} \sum_{x \in P \setminus \{p\}} [p,x]([p,x]-1) = d \quad \text{for all } p \in P$$

(21) $[p,q] = [p',q'] \implies w(p,q) = w(p',q')$
for all $p,q,p',q' \in P$, $p \neq q$, $p' \neq q'$.

Hence for any design D

(22) $2 \cdot l(p) = w(p,p) - \sum_{q \in P \setminus \{p\}} [p,q] \qquad (p \in P)$

and, if D is a $PSI(\alpha,r)$

(23) $2 \cdot l(p) = w(p,p) - \alpha + r \qquad (p \in P)$.

Definition: A design $D = (P, B, I)$ is called *linear strongly regular (LSR) for* a_1, a_2, a_3 $(\in \mathbb{R})$, if
$w(p,q) = a_1 + a_2[p,q]$, $w(p,p) = a_3$
for all $p,q \in P$, $p \neq q$.

Hence by (16) and (23) we have

(24) If D is a regular point stable design, then
D is LSR iff $Mg(D)$ is LSR .

From (19) we obtain

Lemma 1: Let D be a design and H the adjacency matrix of Mg(D), the D is LSR for a_1, a_2, a_3 iff $H^2 - a_2H + (a_1 - a_3)E = a_1J$.

Theorem 1: Let D be a connected PSI(α, r), then

a) D is strong iff Mg(D) is LSR,

b) If D is strong, then Mg(D) is a strongly regular multigraph.

Proof: a) (i) If Mg(D) is LSR, then by (16) and Lemma 1 $H^2 - a_2H + (a_1 - a_3)E = a_1J$. Since $N = H + rE$ we have $Pr(D) \leq 2$, hence D is strong.

(ii) If D is strong, then $Pr(D) \leq 2$. If $Pr(D) = 1$, then D is an (r, λ)-design, hence D is LSR. If $Pr(D) = 2$, then $|\mathrm{spec}N| = 3$, $\mathrm{spec}N = \{\alpha, \rho_1, \rho_2\}$ (say), hence the connection matrix N satisfies the Hoffman-equation

(25) $(N - \rho_1E)(N - \rho_2E) = \xi J$, where

(26) $\xi = (\alpha - \rho_1)(\alpha - \rho_2)v^{-1}$.

Substituting (17) in (25) we obtain

(27) $(H - (\rho_1 - r)E)(H - (\rho_2 - r)E) = \xi J$,

hence by (19)

(28) $w(p,q) = \xi + (\rho_1 + \rho_2 - 2r)[p,q] \qquad (p, q \in P,\ p \neq q)$,

(29) $w(p,p) = \xi - (\rho_1 - r)(\rho_2 - r) \qquad (p \in P)$.

b) Mg(D) is LSR by a), regular by (18) and has constant loop degree d, where $2d = \xi - (\rho_1 - r)(\rho_2 - r) - \alpha + r$ by (23,29).

As an immediate consequence of Theorem 1 we obtain

Corollary 1: a) If D is partial geometric, then Mg(D) is strongly regular. (BOSE, SHRIKHANDE, SINGHI [3])

b) If D is a connected regular multigraph, then D is LSR iff D is strong.

c) If D is a connected regular graph, then D is strongly regular iff D is strong.

4. EIGENVALUES EXPRESSED IN GEOMETRICAL TERMS

The following theorem expresses the eigenvalues of a strong design in geometrical terms.

Theorem 2: Let $D = (\mathcal{P}, \mathcal{B}, I)$ be a strong PSI(α,r) with v points, connection matrix N, spec$N = \{\alpha,\rho_1,\rho_2\}$, $\alpha > \rho_1 > \rho_2$, $\lambda_1, \lambda_2 \in \{[p,q] \mid p,q \in \mathcal{P}, p \neq q\}$, $\lambda_1 > \lambda_2$ and for $i \in \{1,2\}$

(30) $$w_i = \xi + (\rho_1 + \rho_2 - 2r)\lambda_i \quad (= w(p,q), \text{ if } [p,q] = \lambda_i, p \neq q).$$

Then ρ_1, ρ_2 may be expressed in terms of $v,\alpha,r,\lambda_1,\lambda_2,w_1,w_2$:

(31) $$\rho_1 + \rho_2 = \frac{w_1 - w_2}{\lambda_1 - \lambda_2} + 2r \qquad (= c_1 \text{ say })$$

(32) $$\rho_1 \cdot \rho_2 = \alpha\frac{w_1 - w_2}{\lambda_1 - \lambda_2} + v\,\frac{\lambda_1 w_2 - \lambda_2 w_1}{\lambda_1 - \lambda_2} - \alpha(\alpha - 2r) \qquad (= c_2)$$

$$\rho_{1,2} = \frac{c_1}{2} \pm \frac{1}{2}\sqrt{c_1^2 - 4c_2}\ .$$

Proof: Calculate $w_1 - w_2$ to obtain (31). Substitute (26) and (31) in (30) to get (32).

Example: $D = (Z_6, \{\{0,1,3\} + x \mid x \in Z_6\}, \in)$ (add mod 6)

$v=6$, $\alpha=9$, $r=3$, $\lambda_1=2$, $\lambda_2=1$, $w_1=4$, $w_2=6$, hence $\rho_1=3$, $\rho_2=1$.

Note, that D is a divisible design, hence ρ_1, ρ_2 may be calculated shorter by $\rho_1 = \alpha - v\lambda_2 = 3$, $\rho_2 = r - \lambda_1 = 1$.

Remark 1: If D is a strong design with spec$N = \{\alpha,\rho,0\}$, $\alpha > \rho \geq 0$, then D is semi partial geometric and ρ may be determined also by counting p-B-paths, namely

$$\rho = (NA)_{p,B} - (NA)_{q,B}$$

for all $p,q \in \mathcal{P}$, $B \in \mathcal{B}$ with $p\ I\ B$, $q \not{I}\ B$ (cf.[7]).

Remark 2: The author proved in [7], that a strong design $D = (\mathcal{P}, \mathcal{B}, I)$ with exactly two connection numbers λ_1, λ_2 $(\lambda_1 > \lambda_2)$ has a strongly regular *point graph*

$Pg(D) = (\mathcal{P}, \{\{p,q\} \mid p,q \in \mathcal{P}, p \neq q, [p,q] = \lambda_1\}, \in)$.

This theorem contains the result of NEUMAIER [6], that the

point graph of a partial geometric design is strongly regular, and this result contains the well-known fact that the following graphs are strongly regular (The point graph of the dual of D is called the *block graph of D*):

(i) The block graph of a partial geometric design with two intersection numbers,

(ii) the point graph and the block graph of a partial geometry,

(iii) the block graph of a 2-design with two intersection numbers,

(iv) the block graph of a 2-(v,k,1)-design.

5. THE PLACE OF Mg(D) IN THE RANK CLASSIFICATION SCHEME

In the following theorem we determine for any connected regular point stable design D the "place" of the multigraph $D_1 = Mg(D)$ in the rank classification scheme by calculating s_1, $R(D_1)$, $Pr(D_1)$ and checking, whether the connection matrix N_1 resp. the intersection matrix M_1 of D_1 is singular.

Theorem 3: Let D be a connected $PSI(\alpha,r)$, A_1 the incidence matrix of the multigraph $D_1 = Mg(D)$, $N_1 = A_1A_1^T$, $M_1 = A_1^TA_1$.

a) D_1 is a connected regular multigraph with $N_1 = N + (\alpha - 2r)E$, $R(D_1) = s_1 - 1$ $(s_1 = |specN_1 \setminus \{0\}|)$, $Pr(D_1) = Pr(D)$.

b) If $2r < \alpha$, then $detN_1 > 0$, $detM_1 = 0$,
$s_1 = s$, if $detN > 0$, $s_1 = s + 1$, if $detN = 0$.

c) If $2r = \alpha$, then $N_1 = N$ and
$detM_1 > 0$ iff D_1 is a circle of odd length.

Remarks: (i) In case c) D_1 is a circle of odd length iff $D \simeq D_1$ or $D \simeq (\{1,2,3\},\{ \{1\},\{2\},\{3\},\{1,2,3\} \}, \in)$.

(ii) If $\alpha < 2r$, then for each $p \in P$ there is a $B \in \mathcal{B}$ with $p\,I\,B$ and $[B] = 1$. Removing spreads of blocks of length 1 we get a $PSI(\alpha',r')$ D' with $Mg(D') = Mg(D)$ and $\alpha' \geq 2r'$, hence the case b) or c) of Theorem 3 applies, if $\alpha' < \sigma'$. Otherwise D contains only blocks of length 0,1,v.

(iii) Hence the intersection matrix M_1 of the multigraph D_1 of a connected $PSI(\alpha,r)$ is singular, unless D_1 is a circle of odd length or D has only two points, one block of length 2 and $2(r-1) \geq 2$ blocks of length 1.

Proof: a) Clearly D_1 is a connected regular multigraph, hence $R(D_1) = s_1 - 1$ by (9). Since D_1 is $(\alpha-r)$-regular, we get $N_1 = N + (\alpha-2r)E$, hence $Pr(D_1) = Pr(D)$.

b) Let $2r < \alpha$ and ρ be the minimal eigenvalue of N. Then the minimal eigenvalue of N_1 is $\rho + \alpha - 2r > 0$, hence $\det N_1 > 0$. Since $\det N_1 > 0$ we have $v \leq b_1$ = the number of edges in D_1 and $v = b_1$ iff $\det M_1 > 0$. Assume $v = b_1$, then D_1 is a connected regular square graph, hence $D_1 \simeq C_v$, the circle with v points, where $v \geq 3$ since $\alpha < \sigma$ (cf.(1)). Hence $D \simeq D_1 \simeq C_v$, contradicting $2r < \alpha$. Hence $\det M_1 = 0$. Since $\det N_1 > 0$ we have $s_1 = |\mathrm{spec} N_1| = |\mathrm{spec} N| \in \{s, s+1\}$ and $|\mathrm{spec} N| = s$ iff $\det N > 0$.

c) Let $2r = \alpha$, then $N_1 = N$.

(i) It is well-known (cf. COLLATZ, SINOGOWITZ [4]), that if D_1 is a circle C_v $(v \geq 3)$, then $\det M_1 > 0$ iff v is odd.

(ii) Let $\det M_1 > 0$. Then $b_1 \leq v$ and $b_1 = v$ iff $\det N_1 > 0$. If $b_1 = v$, then (as in b)) $D_1 \simeq C_v$ $(v \geq 3)$. By (i) v is odd. If $b_1 < v$, then D_1 must be a tree, since the connected multigraph D_1 contains a spanning tree D_2 with edgenumber $b_2 \leq b_1 \leq v-1 = b_2$, hence $D_1 = D_2$. Therefore $D \simeq D_1$ is a tree with only two points (since D is a $PSI(\alpha,r)$), contradicting $\alpha < \sigma$.

REFERENCES

[1] BOSE, R.C. Strongly regular graphs, partial geometries and partially balanced designs; Pac.J.Math. 13, 389-419, 1963.

[2] BOSE, R.C., BRIDGES, W.G., SHRIKHANDE, M.S. A characterization of partial geometric designs; Discrete Math. 16, 1-7, 1976.

[3] BOSE, R.C., SHRIKHANDE, S.S., SINGHI, N.M. Edge regular multigraphs and partial geometric designs with an application to the embedding of quasi-residual designs; Colloquio Internazionale sulle Teorie Combinatorie, Tomo I, 49-81, Accademia Nazionale dei Lincei, Roma 1976.

[4] COLLATZ, L., SINOGOWITZ, U. Spektren endlicher Graphen; Abh.Math.Sem.Hamburg 21, 63-77, 1957.

[5] HOFFMAN, A.J. On the polynomial of a graph; Amer.Math.Monthly 70, 30-36, 1963.

[6] NEUMAIER, A. $t\frac{1}{2}$ - designs; J.Comb.Th. A28, 226-248, 1980.

[7] WOLFF, K.E. Punkt-stabile und semi-partial-geometrische Inzidenzstrukturen; Mitt.math.Sem.Giessen,135,1-96, 1978.

[8] WOLFF, K.E. Rank classification of point stable designs; (to appear in Europ.J.Comb.)

[9] WOLFF, K.E. Uniqueness of the rank polynomials of point stable designs; Math.Z. 175, 261-266, 1980.

[10] WOLFF, K.E. Point stable designs; Finite geometries and designs, London Math. Soc. Lecture Note Series 49, 365-369, Cambridge University Press 1981